TETRAHEDRON ORGANIC CHEMISTRY SERIES
Series Editors: J E Baldwin, FRS & P D Magnus, FRS

VOLUME 16

Advanced Problems in Organic

Reaction Mechanisms

Related Pergamon Titles of Interest

BOOKS

Tetrahedron Organic Chemistry Series:
CARRUTHERS: Cycloaddition Reactions in Organic Synthesis
DEROME: Modern NMR Techniques for Chemistry Research
GAWLEY & AUBÉ: Principles of Asymmetric Synthesis
HASSNER & STUMER: Organic Syntheses based on Name Reactions and Unnamed Reactions
PAULMIER: Selenium Reagents & Intermediates in Organic Synthesis
PERLMUTTER: Conjugate Addition Reactions in Organic Synthesis
SESSLER & WEGHORN: Expanded, Contracted & Isomeric Porphyrins
SIMPKINS: Sulphones in Organic Synthesis
WILLIAMS: Synthesis of Optically Active Alpha-Amino Acids*
WONG & WHITESIDES: Enzymes in Synthetic Organic Chemistry

JOURNALS

BIOORGANIC & MEDICINAL CHEMISTRY
BIOORGANIC & MEDICINAL CHEMISTRY LETTERS
TETRAHEDRON
TETRAHEDRON: ASYMMETRY
TETRAHEDRON LETTERS

Full details of all Elsevier Science publications/free specimen copy of any Elsevier Science journal are available on request from your nearest Elsevier Science office

* In Preparation

Advanced Problems in Organic Reaction Mechanisms

ALEXANDER McKILLOP

School of Chemical Sciences, University of East Anglia
Norwich, England, U.K.

PERGAMON

U.K.	Elsevier Science Ltd, The Boulevard, Langford Lane, Kidlington, Oxford OX5 1GB, U.K.
U.S.A.	Elsevier Science Inc., 660 White Plains Road, Tarrytown, New York 10591-5153, U.S.A.
JAPAN	Elsevier Science Japan, Tsunashima Building Annex, 3-20-12 Yushima, Bunko-ku, Tokyo 113, Japan

First Edition 1997

Library of Congress Cataloging in Publication Data

A catalog record for this book is available from the Library of Congress

British Library Cataloguing in Publication Data

A catalogue record for this book is available from the British Library

ISBN 0 08 0432565 Hardcover
ISBN 0 08 0432557 Flexicover

Contents

FOREWORD

The development of qualitative ideas of organic reaction mechanisms and their underpinning by quantitative physical organic studies provide the basis of much of our understanding of the practice of organic chemistry today. The challenge presented by mechanistic problems is an essential part of an organic chemist's life-long education, and yet there are few sources of such problems presently available. In this work, Professor McKillop provides us with an excellent series of such problems, selected from his extensive collection, built up over the years. They will represent a stimulating intellectual challenge for organic chemists at all stages in their career, from students to practicing professionals in industry and academe.

J E Baldwin, FRS

PREFACE

Every student of organic chemistry is taught the subject on a mechanism basis, and almost all research in organic chemistry is mechanism based. During the last few years there have been a number of new books and revised editions of older books on the theory and principles of organic reaction mechanisms, but few contain problems on which the reader can test his or her skills and understanding of the subject. Major undergraduate textbooks do have sets of problems which are useful at the introductory level, but there are no recent specialist books of problems at a more advanced level.

This book is a collection of 300 problems which challenge the user to devise reasonable mechanistic interpretations for sets of experimental observations. Almost all of the problems are taken from the literature of the last twenty years. Each is a separate entity, although similar mechanistic themes occur in several quite different problems. Answers are not given, nor are references to the original literature. The user who fails to solve a particular problem and reaches an appropriate level of frustration should be able, relatively quickly, to locate the original literature from the information given in the problem.

It is a pleasure to acknowledge the secretarial skills of Jo Larwood, who prepared the book in camera ready format, and the many helpful criticisms of Dr A J Boulton, who read the complete manuscript and worked through most of the problems.

A McKillop
University of East Anglia

LIST OF PROBLEMS

1. A Tandem Route to 1,2,3,4-Tetrasubstituted Naphthalenes

There is no easy and/or effective method for the preparation of 1,2,3,4-tetrasubstituted naph-thalenes starting from a simple naphthalene derivative and based on classical substitution methodology. A clever new route, based on the now common concept of "tandem" reactions, is illustrated as follows. Heating of the sulfoxide **1** with acetic anhydride at 120°C in the presence of maleic anhydride gave an adduct **2**, $C_{20}H_{16}O_4S$, in 87% yield as a mixture of diastereomers. Reaction of the adduct **2** with PTSA in THF at 25°C gave the naphthalene derivative **3** in quantitative yield. Use of methyl propiolate in place of maleic anhydride did not result in isolation of an adduct; the product was the tetralone **4** (51%).

Elucidate these transformations, deduce the structure of the adduct **2**, and give mechanisms for the formation of **2**, **3** and **4**.

2. Hydroazulenes by Radical Cyclisation

Recent studies on radical-induced cyclisation reactions have led to a simple, one step method for the preparation of hydroazulenes from appropriately substituted cyclopentanone precursors. Treatment of **1**, for example, with triphenyltin hydride and AIBN in refluxing toluene gave **2** in 79% yield.

Give a mechanism for this transformation.

3.   Cinnolines to Indoles

A Japanese group have been investigating the use of readily accessible dihydrocinnoline derivatives of the type **1** (R = alkyl, aryl, styryl, ethoxy; X = H or methoxy) as novel precursors for the synthesis of other types of heterocycles.   The following synthesis of 2-acetyl-3-cyanoindole is representative of a new general method (22-88% for 10 examples) for the preparation of 2-acyl- and 2-ethoxycarbonyl-3-cyanoindoles from **1** : a mixture of **1** (R = Me, X = H; 1 eq.) and powdered potassium cyanide (2 eq.) was stirred overnight in aqueous DMF at room temperature.   Addition of water to the reaction mixture precipitated 2-acetyl-3-cyanoindole, which was obtained in 54% yield after recrystallisation.

Suggest a mechanism for this transformation.

1

4.   Rearrangement During NMR Studies

Marine algal overgrowths are known to be a contributing factor in the destruction of hard coral. Examination of one such algal overgrowth collected in Okinawa resulted in isolation of a number of interesting new natural products, including the C_{11} dihydroxytrienone nakienone A, **1**. During study of the nmr spectrum of **1** (in CDCl$_3$ solution) an interesting rearrangement was found to occur in the nmr sample to give the hemiacetal **2**.

Suggest a mechanism for the transformation of **1** into **2**.

1   **2**

5. Synthesis of Trifluoromethyl Heterocycles

Trifluoroacetylketenes, easily prepared as such or as their synthetic equivalents by reaction of acyl chlorides RCH_2COCl with trifluoroacetic anhydride in pyridine, have been found to be valuable precursors to a range of trifluoromethyl heterocycles. In contrast to simple ketenes, the highly electrophilic trifluoroacetylketenes react with electron rich olefins such as imines, vinyl ethers and emamines to give the products of formal [4 + 2] cycloaddition. Condensation with N, N-dimethylcyanamide also proceeded smoothly, and with $R = Me(CH_2)_{13}$ the 1,3-oxazin-4-one **1** was obtained in 78% yield. If, however, the reaction with the cyanamide was carried out using trifluoroacetylketene from which excess trifluoroacetic anhydride had *not* been removed prior to addition of the cyanamide, the product obtained on quench of the reaction mixture with methanol was the dihydro-1,3-oxazin-4-one **2a** (60%). **2a** could be converted quantitatively into **2b** by heating in refluxing ethanol for 2 days.

Suggest a mechanism for the formation of **2a** from **1**.

1, R = Me(CH_2)_{13}   **2a**, R^1 = Me   **2b**, R^1 = Et

6. 3-Substituted Chromones from Pyranobenzopyrans

The pyranobenzopyran **1** is readily available and has proved to be a useful starting material for the preparation of a number of 3-substituted chromones. Thus, reaction of **1** with a large excess of an enolisable ketone $RCOCH_3$ in the presence of aqueous acid resulted in smooth conversion into the chromones **2**.

Give a mechanism for this transformation, and outline a simple synthesis of **1**.

1   **2**

R = Me, Ph, 2-MeOC_6H_4

7.   Acid-Catalysed Rearrangements of 1-Arylindoles

Many acid-catalysed rearrangements of 2- and 3-substituted indoles are known. For example, 3-alkyl and 3-aryl groups migrate to the 2-position; 2-acyl groups migrate under the same conditions to the 3-position. In some cases the indole ring undergoes cleavage followed by some kind of recyclisation. Similar types of reaction are rare with 1-substituted indoles, but a recent report describes an interesting and potentially useful example of such a process. Thus, heating of 1-arylindoles **1** with PPA at temperatures between 75 and 115°C for periods of 25 to 150 hours gave the 5*H*-dibenz[*b, f*]azepines **2** in yields of 8-65%. The highest yield of **2** was obtained with **1** (R = 4-Me), and no rearrangement to products **2** was observed with **1** (R = 4-NO$_2$) or **1** (R = 3-CF$_3$).

Suggest a mechanism for the **1** → **2** transformation consistent with these observations.

1   **2**

8.   An Unusual Deprotection of an Aryl Ester

Conversion of 2',4',6'-trihydroxyacetophenone to the tricarbonate proceeded smoothly in 87% yield on treatment with 3.3 equivalents of methyl chloroformate and triethylamine in THF at 0°C. Treatment of this tricarbonate with 4 equivalents of sodium borohydride in a 1:1 THF/water mixture at 0°C to ambient temperature gave the phenol **1** in 83% yield.

Suggest a mechanism to account for this transformation.

1

9.   3,1-Benzoxathiin Formation

Give a mechanistic explanation for the observation that heating of the sulfoxide **1** with 1.2 equivalents of PTSA in xylene for 50 minutes results in formation of the 3,1-benzoxathiin **2** (53%).

1   **2**

10.   Arylacetone Synthesis by Carroll-Type Rearrangement

The Carroll rearrangement, a variation on the ester Claisen rearrangement, is a useful method for the preparation of γ, δ-unsaturated ketones from allyl acetoacetates, and has been adapted to provide a method for the synthesis of a number of specific arylacetones. Thus, treatment of the *p*-quinol **1** with diketene and a catalytic amount of DMAP at room temperature gave a 72% yield of the arylacetone **2** together with 5% of the benzofuran **3**.

Suggest a mechanism for the formation of **2**.

1   **2**   **3**

R = PhC≡ C

11.   Attempted Knoevenagel Reaction Gives Mannich-Type Products

Attempted Knoevenagel condensation of the coumarin aldehydes **1** using piperidine as base led in a number of cases to formation of significant amounts of Mannich-type products **2** ($R^2, R^3 =$

$(CH_2)_5$). A more detailed study of this process showed that treatment of **1** with a variety of secondary amines R^2R^3NH in hot DMF or heptane gave products of the type **2** in 24-87% yield. Aldehydes R^1CHO were also formed.

Give a mechanistic explanation for these transformations.

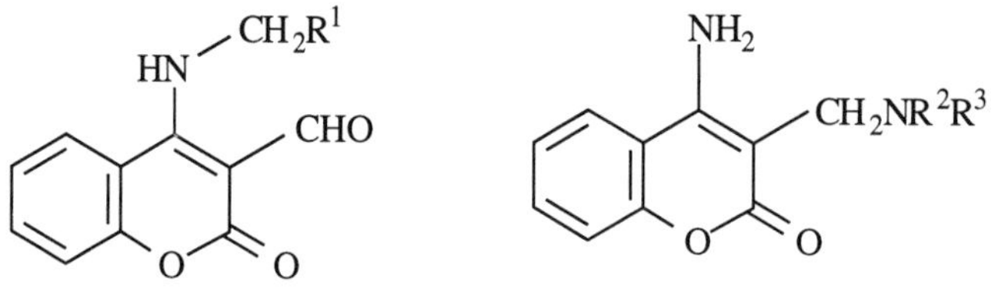

1

2

12.   Pyridinium Salt Rearrangements

Highly substituted pyridinium salts of type **1** are easily accessible by Hantzsch-type synthesis. They are valuable products in that they can often be transformed easily into otherwise difficultly accessible, highly substituted benzenes or pyridines by base-catalysed rearrangement. For example, treatment of the salt **2** with ethanolic sodium hydroxide at room temperature for one hour gives 2,4-diacetyl-*N*,5-dimethylaniline **3** in 85% yield, while 3-acetyl-5-cyano-6-methyl-2-methylamino-4-phenylpyridine **5** is obtained in 88% yield from the salt **4** under the same conditions.

Give mechanisms for the transformations **2 → 3** and **4 → 5**.

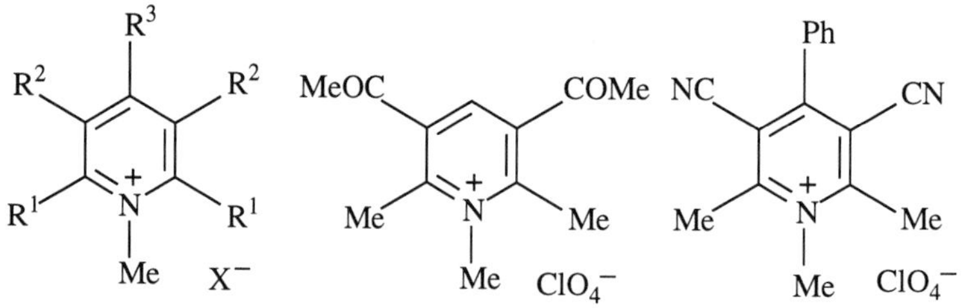

1   2   4

13. A Pyridazine from a Thiophene Dioxide

A product **A**, $C_{28}H_{30}N_6O_4$, was obtained in 87% yield when a mixture of 3,4-di-t-butylthiophene 1,1-dioxide and 4-phenyl-1,2,4-triazoline-3,5-dione (2 eq.) was heated under reflux in toluene for 5 hours. Hydrolysis of **A** with methanolic potassium hydroxide at room temperature gave an 80% yield of 4,5-di-t-butylpyridazine.

Give a structure for **A** and a mechanism for its transformation into the pyridazine.

14. Biomimetic Synthesis of Litebamine

Treatment of the isoquinolinium methiodide **1** with sodium methoxide in methanol for two hours at reflux temperature gave the tetrahydroisoquinoline **2** in quantitative yield, and *not* the anticipated benzazepine **3**. This unexpected result was then exploited as follows for a "biomimetic" synthesis of the alkaloid litebamine. Thus, quaternisation of (S)-O,O-diacetylboldine **4** with iodoacetonitrile in methylene chloride (41%) followed by treatment of the salt with NaOMe/MeOH gave litebamine **5** (69%).

Give mechanisms for the transformations of **1** into **2** and of **4** into **5**, and explain on a mechanistic basis why **3** was initially expected to be the product obtained from treatment of **1** with base.

1

2

3

4

5

15.   Acid-Catalysed Degradation of *N*-Nitroso-2,3-didehydromorpholine

N-Nitrosodiethanolamine is a potent, widespread animal carcinogen, metabolic oxidation of which by alcohol dehydrogenase is known to generate *N*-nitroso-2-hydroxymorpholine, which is mutagenic but not carcinogenic. In a study aimed at probing potential biologically significant transformations of *N*-nitroso-2-hydroxymorpholine it was decided to investigate its possible reversible conversion to *N*-nitroso-2,3-didehydromorpholine. The latter compound was synthesised, but was found *not* to react with neutral water at room temperature. It was slowly decomposed by aqueous sodium hydroxide solution, but reacted smoothly with at least one equivalent of hydrochloric acid in water at room temperature to give *N*-(2-hydroxyethyl)-2-oximinoethanamide in 89% yield.

Give a mechanism for the formation of this latter compound.

16.   A New Route to 4-Amino-3-arylcinnolines

A new, efficient and general synthesis of 4-amino-3-arylcinnolines has been described recently, starting from arylhydrazines and aryl trifluoromethyl ketones, and the following is a typical example. Addition of a THF solution of the hydrazone formed from phenylhydrazine and 2-trifluoroacetylthiophene to 5 equivalents of KHMDS in THF at -78°C, then slow warming of the mixture to room temperature and stirring at room temperature for several hours, gave 4-amino-3-(2-thienyl)cinnoline in 68% yield after work-up (quench with ether, then a wash with brine).

Suggest a mechanism for this transformation.

17.   Base-Induced Cyclisations of *o*-Ethynylaryl-Substituted Benzyl Alcohols

In a study of the base-induced cyclisations of *o*-ethynylaryl-substituted benzyl alcohols it was found that substrates **1a-c** reacted smoothly and exclusively by 5-*exo-dig* ring closure to give the isobenzofuran derivatives **2a-c**. Thus, treatment of **1a** with KOH/MeOH gave **2a** in 95% yield; **2b** and **2c** were obtained in quantitative yield by treatment of **1b** and **1c** with NaH in THF.

1

a, R¹ = R² = H   **b, R¹ = H; R² = OMe**   **c, R¹ = H; R² = NO₂**

Under the same conditions of NaH/THF, the ester **3** gave 1*H*-2-benzopyran derivative **5** in 60% yield, apparently by 6-*endo-dig* ring closure. Closer study of this latter transformation, however, revealed that the initial product of base-induced cyclisation was in fact the isobenzofuran **4**, which was extremely labile, and that **5** was formed from **4** by acid-catalysed rearrangement during work-up of the reaction mixture.

Suggest a mechanism for the conversion of **4** into **5**.

3   **4**   **5**

18.   Trifluoromethylpyrroles from Trifluoromethyloxazolones

Transformation of a readily available heterocycle into a less readily available heterocycle is a time-honoured and valuable strategy in synthesis. In a recent example, a versatile and high yielding synthesis of trifluoromethylpyrroles was described, starting from trifluoromethyloxazolones. Typically, 2-trifluoromethyl-4-phenyl-5(2*H*)-oxazolone was mixed with α-chloroacrylonitrile in methylene chloride at room temperature, triethylamine was added, and the mixture stirred. Standard work-up gave 3-cyano-2-phenyl-5-trifluoromethylpyrrole in 70% yield.

Suggest a mechanism for this ring transformation.

19.   The Hooker Oxidation

The Hooker oxidation, a classical and historic example of which is the conversion of lapachol **1** to the lower homologue **2**, has been described as "......one of the more remarkable reactions in organic chemistry." It was also pointed out that "It is suprising that its synthetic potential has not been realised......"

$$KMnO_4/NaOH$$
or
1) H_2O_2/Na_2CO_3
2) $CuSO_4/NaOH$

1   **2**

The Hooker oxidation was first described in 1936, a mechanism was proposed in 1948, and a recent labelling study has provided data in support of the 1948 mechanism. Thus, Hooker oxidation of the naphthoquinone **3** gave **4** in 72% yield.

Deduce a mechanism for the Hooker oxidation which is consistent with the results of the labelling study.

3   **4**

$* = {}^{13}C$ enrichment

20.   Synthesis of 1,3-Disubstituted Pyrroles

Russian workers have recently described an interesting synthesis of 1,3-disubstituted pyrroles which consists of the following 3-step procedure, illustrated for the preparation of 1-benzyl-3-ethylpyrrole (from which 3-ethylpyrrole is easily available by Na/NH$_3$ reductive removal of the benzyl group):

(1)   1,1-Diethoxyhex-3-ene was treated with one equivalent of bromine in pentane at -10°C.

(2)   The crude product from (1) was stirred with formic acid at room temperature to give **A**.

(3)   The product from (2) was heated at reflux with benzylamine.

Elucidate the reaction scheme, identify **A** and suggest a mechanism for the conversion of **A** into 1-benzyl-3-ethylpyrrole.

21.   Grob Fragmentation of a 1-Azaadamantane

Treatment of the readily accessible 1-azaadamantane derivative **1** with thionyl chloride gave the mono-*gem*-dichloro derivative in 92% yield. Refluxing of a solution of the latter in aqueous ethanol containing triethylamine (3 eq.) gave 5-chloro-2,4,6-trimethylresorcinol in 62% yield in what has been claimed as the first example of a triple Grob fragmentation reaction.

Give a mechanism for the formation of the resorcinol.

1

22.   Selective Substitution of 3-Methoxypyrazine *N*-Oxide

Azine oxides are versatile starting materials for heterocyclic synthesis and are frequently used for regioselective ring substitution reactions, most of which proceed with loss of the oxide substituent. Occasionally some unusual selectivities are observed. For example, treatment of 3-methoxypyrazine *N*-oxide with equimolar amounts of diethylcarbamoyl chloride and 4-methoxytoluene-α-thiol in refluxing acetonitrile gave 2-methoxy-6-(4-methoxybenzylthio)-pyrazine as the sole product in 60% yield.

Suggest a mechanism to account for the regioselectivity of substitution.

23.   Reactions of Ethacrylate Esters with NO₂BF₄

Treatment of ethacrylate esters **1** with nitronium tetrafluoroborate in acetonitrile has been shown to give cyclopropanes **2** and the products of allylic nitration **3**. Formation of **2** was postulated to proceed via an α-carbonyl cation. In an attempt to obtain evidence for the possible intermediacy of α-carbonyl cations in these reactions in terms of Wagner-Meerwein derived products, the more highly substituted substrates **4a, b** were subjected to the same reaction conditions of NO₂BF₄/MeCN followed by aqueous work-up. This gave **5a, b** and **6a, b** as shown.

Suggest mechanisms for the formation of products **5** and **6**.

4a, R = H
b, R = Me

5a, R = H, 17%
b, R = Me, 8%

6a, R = H, 37%
b, R = Me, 9%

24.   A Furan Synthesis

Treatment of methyl vinyl ketone with sodium benzyloxide (1 eq.) followed by bromine (1 eq.) gave the expected alkoxy bromo ketone. Reaction of the latter with DBU in benzene at room temperature gave a 70% yield of 5-acetyl-3-benzyloxymethyl-2-methylfuran.

Suggest a mechanism for this transformation.

25. Dimerisation of Vinca Alkaloids

During studies on the synthesis on vinca alkaloids and related compounds it was found that (-)-criocerine **1** and a number of analogues underwent smooth dimerisation when dissolved in acetic acid at room temperature for 24 hours. Thus, **1** gave **2** in 86% yield.

Give a mechanism for this transformation.

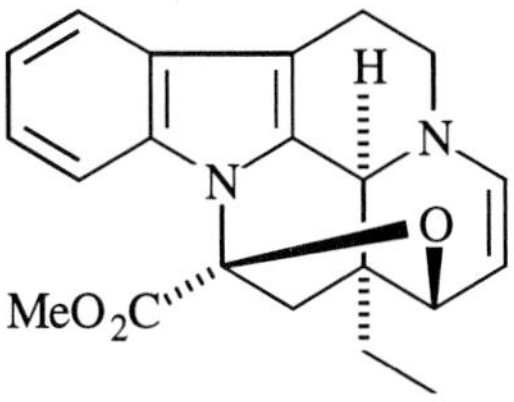

1

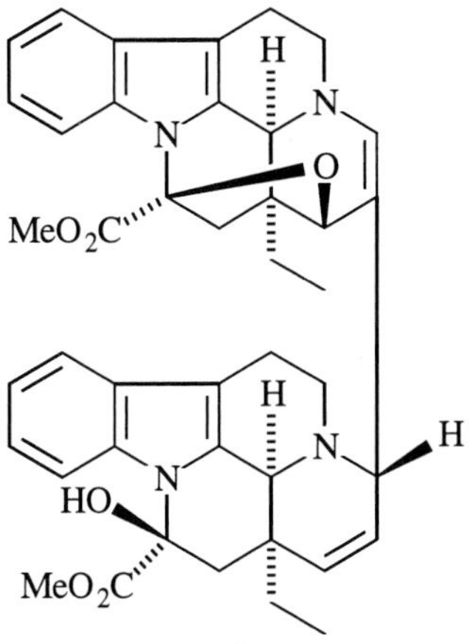

2

26. Nitration of a Quinoline Derivative

Nitration of quinolines in the benzo ring is seldom a problem, especially if the benzo ring contains an electron donating group. For some undisclosed reason, 6-methoxy-4-methylquinoline was subjected to fuming nitric acid for 3 days at room temperature, and this gave a 20% yield of the lactone **1**.

Suggest a mechanism for the formation of **1**.

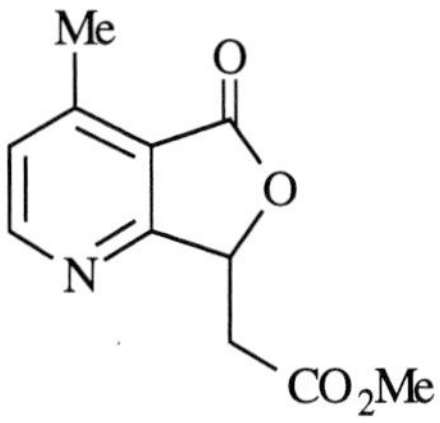

1

27.   Synthesis of Binaphthyldiquinones

A number of "binaphthyldiquinones" have recently been found as natural products, and Russian workers have described a very simple, one-step method for preparation of the basic system. Thus, diquinone **1a** was obtained in 40% yield simply by heating a solution of 1,4-naphthoquinone with two equivalents of sodium hydride in THF. Use of 2-chloro- or 2-bromo-1,4-naphthoquinone gave the diquinones **1b** and **1c** in 94 and 92% yield respectively.

Give a mechanism for this dehydrodimerisation reaction.

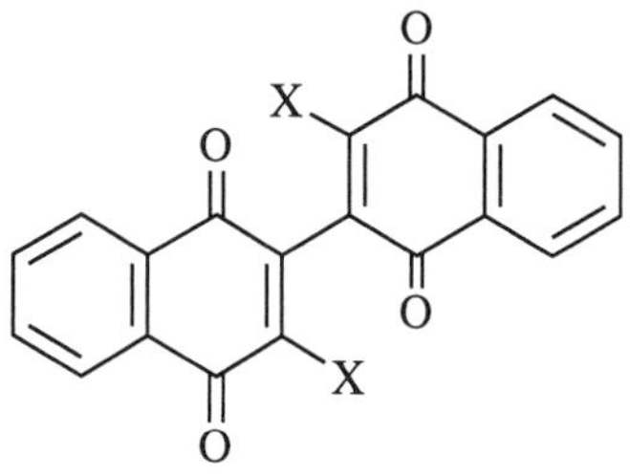

$$1a, \quad X = H$$
$$b, \quad X = Cl$$
$$c, \quad X = Br$$

28.   Heterocyclic Fun with DMAD

Addition of triethylamine to a stirred suspension of the salt **1** and potassium carbonate in toluene at room temperature resulted in formation of a yellow colour. Addition of 2.5 equivalents of DMAD produced a red colour, and the pyrazolo[1,5-*a*]pyridine **2** was obtained in 82% yield.

Give a mechanism for the transformation of **1** into **2**.

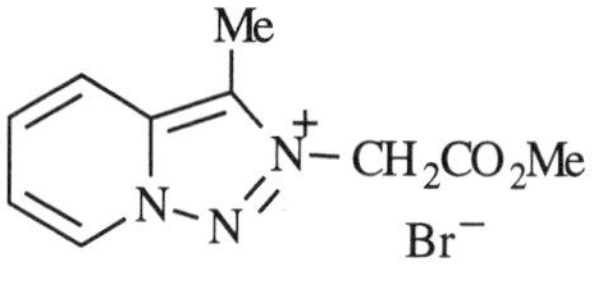

1 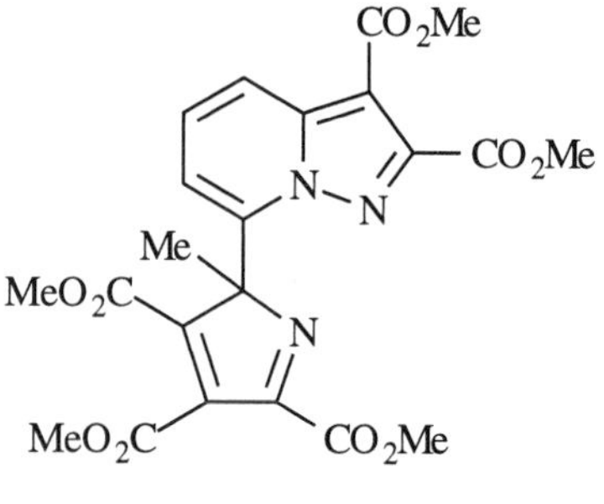   **2**

29. Unexpected Formation of a 1,5-Benzodiazonine

During studies on the synthesis of the alkaloid aaptamine, the *N*-methoxyamide **1** was treated with PIFA (PhI(OCOCF$_3$)$_2$) in chloroform at reflux temperature in the expectation that the product would be **2**. The only product isolated, however, was the 1,5-benzodiazonine derivative **3** (76%).

Suggest a mechanism for the formation of **3**.

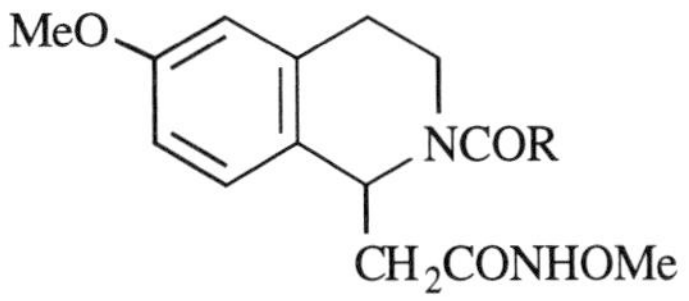

1

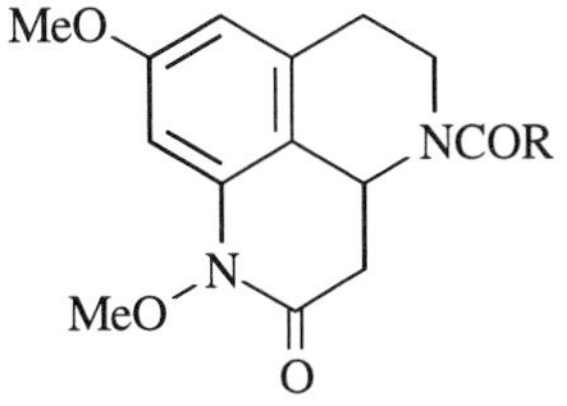

2

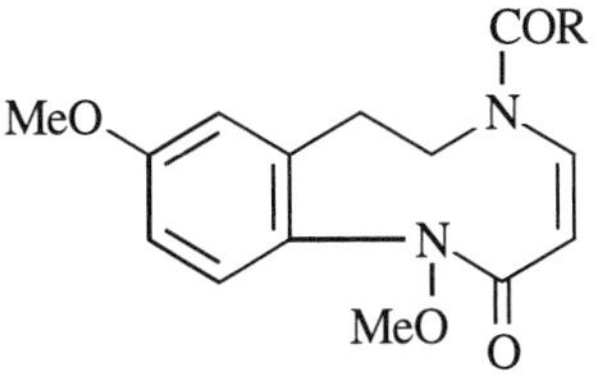

3

30. A Thiopyran to Thiophene Transformation

1*H*-2-Benzothiopyran 2-oxide reacts smoothly with an equimolar amount of acetylacetone in acetic anhydride at 100-110°C to give 1-(diacetylmethyl)-1*H*-2-benzothiopyran in 80% yield. Treatment of 1-phenyl-1*H*-2-benzothiopyran 2-oxide with a three fold excess of acetylacetone in acetic anhydride under the same conditions, however, gave a complex mixture of products, the main component of which (16%) was found to be the red, crystalline benzo[*c*]thiophene derivative **1**.

Suggest a mechanism for the formation of **1**.

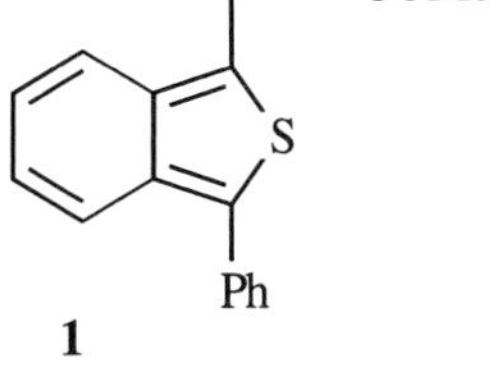

1

31.   Skeletal Rearrangements of a Diterpene

Reaction of methyl pimarate **1** with bromine in THF/H$_2$O in the presence of NaHCO$_3$ for 10 minutes at 0°C gave a mixture of the products **2** (41%), **3** (18%), **4** (5%) and **5** (11%). The 6-6-7 system in **2-5** is the skeleton of the strobane diterpenes, and the **1** → **2-5** process has been claimed to provide "a biomimetic access to strobane derivatives".

Give mechanisms for the transformation of **1** into **2-5**.

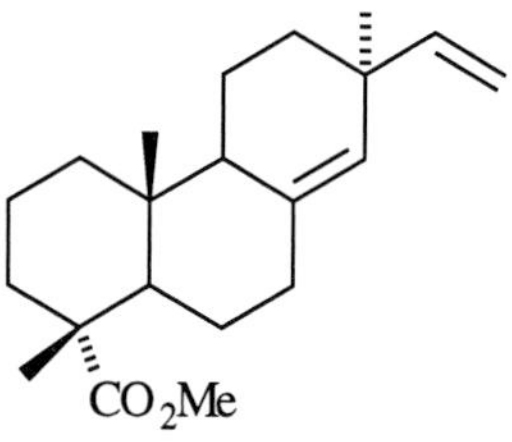

1

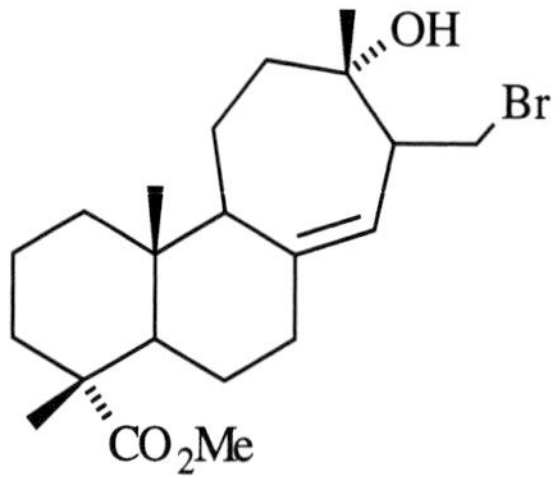

2

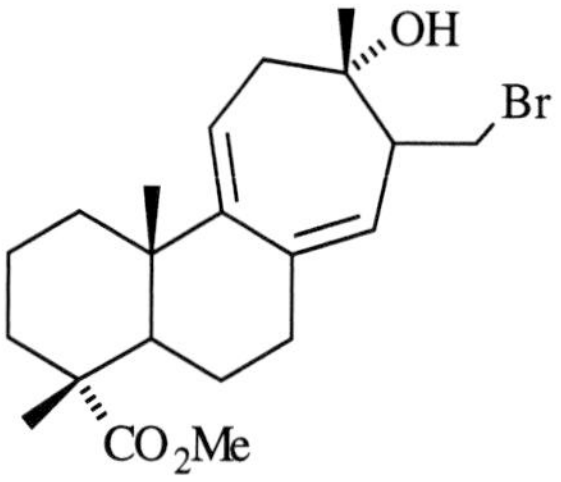

3

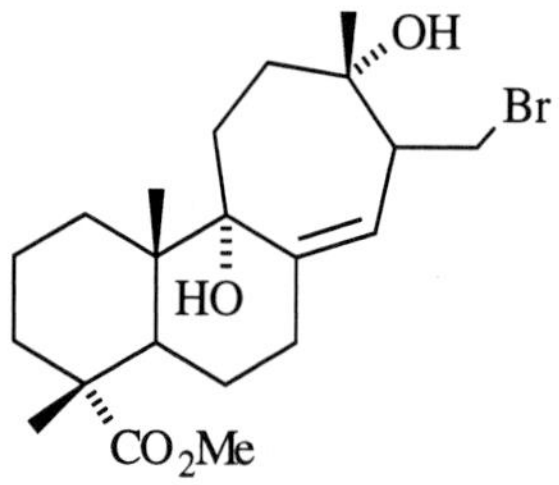

4

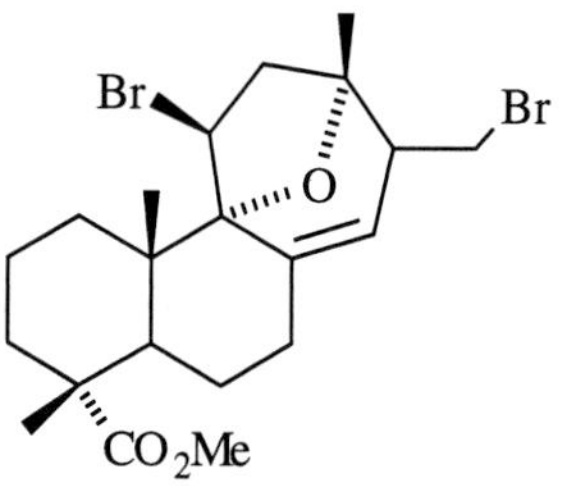

5

32.   Reissert Reactions of Quinoxaline *N*-Oxides

Studies of the reactions of quinoxaline *N*-oxides under Reissert reaction conditions have led to some very interesting and unusual results. Thus, treatment of quinoxaline *N*-oxide with PhCOCl/KCN in methanol or water under standard Reissert conditions gave 6-chloroquinoxaline as the major product (*ca.* 45%), and only small amounts of the desired 2-cyanoquinoxaline. Use of 3 equivalents of TMSCN in place of KCN and dichloromethane as solvent, however, gave 2-cyanoquinoxaline in 72% yield. When 2,3-diphenylquinoxaline *N*-oxide was treated with 1 equivalent of PhCOCl in the presence of 3 equivalents of either KCN or TMSCN a mixture of products was always obtained irrespective of the solvent used. The most interesting of these products was the ring cleaved compound **1**.

Elucidate these various transformations and suggest a mechanism for the formation of **1**.

1

33.   Unexpected Results in Directed Metallation

An unexpected reaction was encountered during attempts to use the directed metallation approach for *peri*-functionalisation of polycyclic aromatic hydrocarbons. Thus, reduction of the ketone **1** with borohydride followed by reaction of the product with pivaloyl chloride in pyridine in the presence of DMAP gave the expected pivalate ester. Lithiation of this ester with 2-3 equivalents of LDA in THF at -50°C followed by TMSCl quench at 0°C gave a product **2** the nmr spectrum of which showed not only the presence of the t-butyl and TMS groups, but also the *peri*-H-6 signal. The signal for H-7 was absent. Repetition of the lithiation reaction, but work-up of the reaction mixture with aqueous ammonium chloride rather than quench with TMSCl gave a 69% yield of a polar product **3**. The nmr spectrum of this again showed no signal for H-7, but there was a new signal, a singlet, at 5.33 ppm which was exchangeable with MeOH-d$_4$. The H-6 resonance was shifted upfield from 8.06 ppm in **2** to 7.64 ppm. The same product **3** could be prepared independently by (i) condensation of the ketone **1** with the anion derived from the TMS ether of the cyanohydrin of pivalaldehyde, and (ii) treatment of the product thus obtained with TBAF.

Deduce the structures of **2** and **3** and suggest a mechanism for their formation.

1

34.   1,4-Dioxene in Synthesis

A French worker has recently been investigating the utility of 1,4-dioxene (more correct is : 2,3-dihydro-1,4-dioxin) in organic synthesis and discovered some interesting and potentially very useful transformations.   For example, lithiation of 1,4-dioxene and addition of 2,2,6-trimethylcyclohexanone proceeded normally to give, after dehydration of the initially formed adduct, the expected diene **A**. Oxidation of **A** with MCPBA in methanol at 0°C gave **B**, $C_{14}H_{24}O_4$, in 70% yield as a mixture of diastereomers. Low temperature reaction of **B** with allyltrimethylsilane catalysed by $EtAlCl_2$ gave the protected α-hydroxy ketone **1** in 85% yield.

Outline the overall reaction scheme and suggest a mechanism for the conversion of **B** into **1**.

1

35.   Collins Oxidation of Phenylethanols

Unexpected reactions can often occur during attempts to carry out what appear to be quite trivial transformations, as illustrated by the following example.   A number of 4-substituted phenylacetaldehydes were required as intermediates for the preparation of a series of compounds for study as potential serine protease inhibitors. These were to be prepared by standard oxidation of the corresponding phenylethanols with Collins reagent (CrO$_3$/pyridine/CH$_2$Cl$_2$). Oxidation of the phenylethanols **1a** and **1b** did indeed give the corresponding phenylacetaldehydes in "good to excellent" yield, but small amounts (<5%) of 4-chloro- and 4-methoxybenzaldehyde were also formed. When **1c** was treated with Collins reagent under the same conditions as used for **1a** and **1b**, however, no 4-nitrophenylacetaldehyde was obtained; the only product was 4-nitrobenzaldehyde (~30%), together with *ca.* 70% of unchanged starting material. There was no reaction of Collins reagent with **1d**.

Give a mechanistic explanation for these observations.

CH$_2$CH$_2$OH

R

1a, R = Cl

b, R = OMe

c, R = NO$_2$

d, R = NMe$_2$

36.   Allene Sulfoxide Rearrangement

Ethynylation of the ketone **1** and treatment of the propargyl alcohol thus obtained with phenylsulfenyl chloride in the presence of triethylamine gave the expected allene sulfoxide. Treatment of this sulfoxide with aluminium trichloride in benzene gave the sulfide **2** in 15% yield.

Elucidate the reaction sequence and give a mechanism for the formation of **2**.

Me

Me

SPh

1   **2**

37.   A Failed "Pinacol-Type" Rearrangement

The diterpene derivative **1** was treated with $BF_3.OEt_2$ and Ac_2O in an attempt to induce a "pinacol-type" rearrangement leading to a ring B-homo derivative. No such reaction was observed. Instead, the only product which could be isolated (33%) was shown to be **2**, formed by "a profound backbone rearrangement" of **1**.

Suggest a mechanism for this rearrangement.

1

2

38.   Allylation of 1,4-Benzoquinones

Simple, direct and regioselective allylation of 1,4-benzoquinones is an important objective in the synthesis of many isoprenoid quinones such as plastoquinone-1 and vitamin K_1, which play important roles in cellular metabolism and photosynthesis. In a new procedure where the use of a strongly coordinating solvent was found to play an important role, it was shown that reaction of 3 equivalents of (3-methyl-2-butenyl)trifluorosilane with one equivalent of 2,3-dimethyl-1,4-benzoquinone in formamide in the presence of 5 equivalents of $FeCl_3.6H_2O$ for 23 hours at 40°C gave plastoquinone-1 **1**, in 90% yield.

Give a mechanistic interpretation for this transformation which takes into account the probable role of the solvent.

1

39.   Benzopyrene Synthesis - By Accident

During studies on the synthesis of trihydroxybenzopyrene derivatives it was found that treatment of the pentacycle **1** with HI/50% H_3PO_2 at 100°C for 2 to 3 minutes resulted in smooth conversion into the trimethoxybenzopyrene **2** in 95% yield.

Suggest a mechanism for this transformation.

1   **2**

40.   Rearrangement of Ketene Acetals

Belgian workers recently described the synthesis of a series of ketene acetals **1** by standard methodology, i.e. treatment of esters $ArCH_2CO_2CH_2SMe$ with LiHMDS/THF at -78°C followed by TBDMSCl at -78 to 20°C. Unexpectedly, these compounds were found to rearrange to *o*-methylthiomethylarylacetic esters after either a few hours in THF at room temperature, or less than one hour at reflux temperature. For example, **1** (R^1 = 4–Cl) gave **2** (R^1 = 4–Cl; R^2 = TBDMS), which on hydrolysis gave the arylacetic acid **2** (R^1 = 4–Cl; R^2 =H) in 76% yield. The (*E*) isomers of **1** rearranged much faster than the (*Z*), and their disappearance followed first order kinetics.

Suggest a mechanism for this unexpected rearrangement.

(*E* : *Z ca.* 9:1)

1   **2**

41.   Nucleophilic Additions to a Heterocyclic *o*-Quinone

"Heterocyclic *o*-quinone cofactors" have been a subject of great interest during the last fifteen years, and PQQ, **1**, in particular has been the objective of intensive study. It is a growth stimulating substance for microorganisms, is nutritionally important for mammals, and is the redox cofactor of an important class of dehydrogenases. Both PQQ, **1**, and the trimethyl ester **2** show very high reactivity towards nucleophiles, and detailed studies of their reactions with water and alcohols have provided valuable information on the possible biological action of PQQ-containing enzymes ("quinoproteins") as alcohol dehydrogenases. Under neutral conditions, PQQ trimethyl ester **2** reacts with methanol to give the C5-hemiacetal, the structure of which was established by X-ray analysis. Under acidic conditions, however, reaction with methanol gave the C4-dimethyl acetal, the structure of which was again confirmed by X-ray analysis.

Give a mechanistic explanation for these observations.

1, R = H
2, R = Me

42.   How Mitomycin C Can Crosslink DNA

Mitomycin C, **1**, is a potent antitumor antibiotic discovered by Japanese scientists in fermentation cultures of *Streptomyces caespitosus*. It has been described as "small, fast and deadly (but very selective)" and has an extraordinary ability to crosslink the complementary strands of the DNA double helix with high efficiency and absolute specificity. It is so lethal that one crosslink per genome is sufficient to cause death of a bacterial cell. Mitomycin C, which is widely used clinically as an antitumor drug, does not react with DNA, but enzymatic reduction of the quinone induces a cascade of transformations which results, ultimately, in formation of the DNA crosslink **2**.

Give a mechanistic interpretation of the transformation of **1** into **2**.

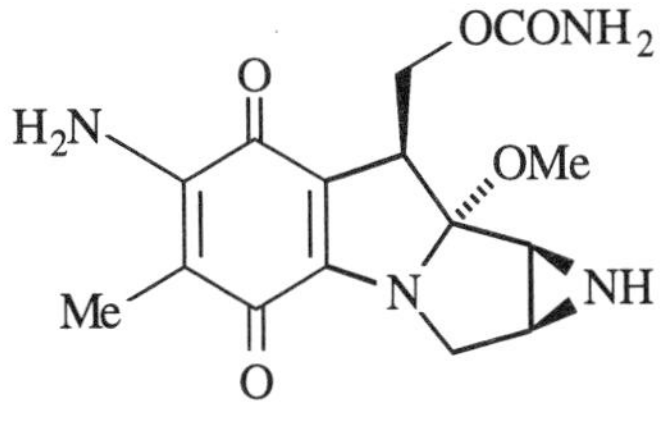

1

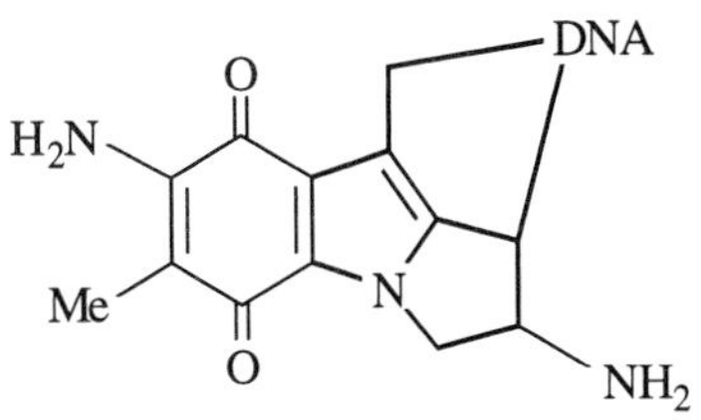

2

43.   Two Carbon Ring Expansions of Cycloalkanones

Reactions of carbocyclic β-keto esters, sulfonium ylides and enamines with activated alkynes such as DMAD are known to result in formation of (n + 2) ring expanded products. In a study of the analogous reactions of carbocyclic β-keto phosphonates, it was found that in the cases of the simple cyclic β-keto phosphonates **1**, ring expansion occurred to give **2** in reasonable yield. Extension of the method to the tetralone **3**, however, led to formation of two products, the "expected" (n + 2) ring expansion product analogous to **2** (37%), and the lactone **4** (29%).

Give mechanisms for the conversion of **1** into **2** and for the formation of **4**.

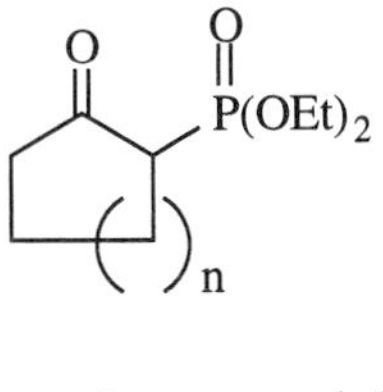

1   n = 1-4

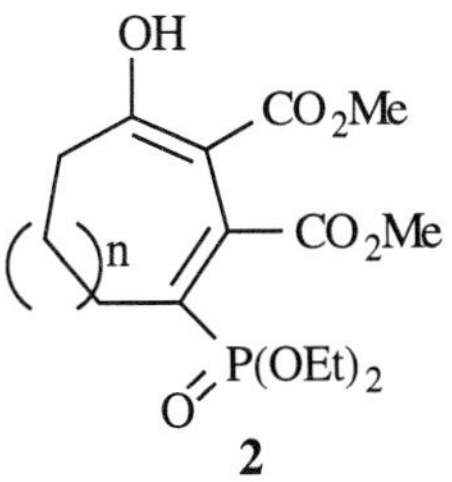

2

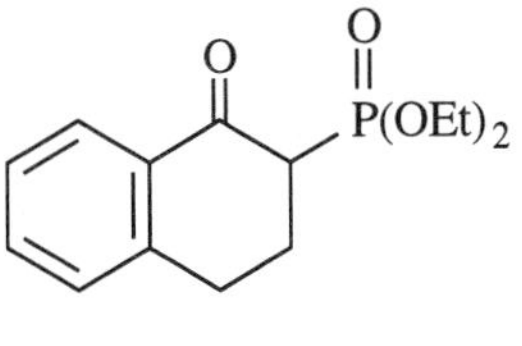

3

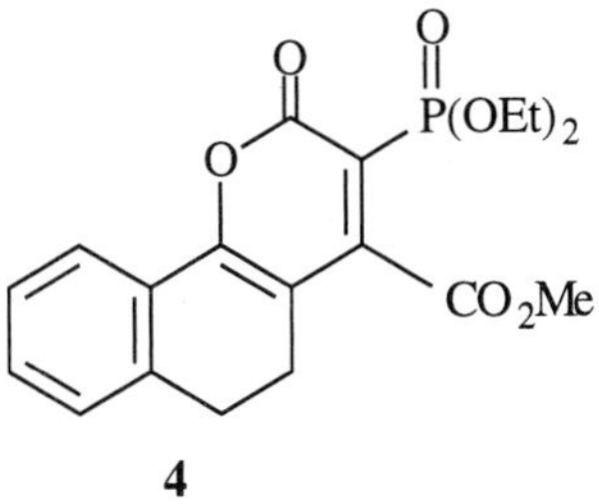

4

44.   Fulgides : Synthesis and Photochromism

Many "fulgides", compounds which undergo facile valence bond tautomerism on irradiation, are of interest in terms of photochromism.   One such compound, 5-dicyanomethylene-4-dicyclopropylmethylene-3-[1-(2,5-dimethyl-3-furyl)ethylidene]tetrahydrofuran-2-one was prepared as a mixture of (E) and (Z) isomers by the following sequence of reactions. Condensation of dicyclopropyl ketone with diethyl succinate using sodium hydride as base gave **A**, $C_{13}H_{18}O_4$, after acidic work-up. Esterification of **A** with EtOH/HCl and condensation of the product with 3-acetyl-2,5-dimethylfuran using LDA/THF at -75°C gave **B**, $C_{21}H_{26}O_5$, as an (E, Z) mixture. Hydrolysis of **B** with KOH/EtOH, acidification, and treatment of the product with acetyl chloride gave **C**, $C_{19}H_{20}O_4$, which was condensed with malononitrile in the presence of diethylamine to give the bis-salt **D**, $[C_{22}H_{20}N_2O_4]^{2-}.2Et_2\overset{+}{N}H_2$. Treatment of **D** with acetyl chloride gave the target fulgide, $C_{22}H_{20}N_2O_3$. Irradiation of a dilute pale yellow solution of the fulgide (solvent not specified) at 350 nm gave a bluish green solution which absorbed at 620 nm; irradiation of this solution at 532 nm caused reversion to the original pale yellow colour.

Elucidate the synthesis of the fulgide and give a mechanistic explanation for its photochromic behaviour.

45.   Benzodiazepines from (R)-(+)-Pulegone

(R)-(+)-Pulegone **1**, a major constituent of the essential oils of *Mentha* species, is a cheap, readily available and valuable synthetic precursor from the chiral pool. When a solution of equimolar amounts of **1** and *o*-phenylenediamine in toluene was heated at reflux for 4 days, the tricyclic benzodiazepine derivative **2** was obtained in 68% yield.   5-Methyl- and 5-chloro-*o*-phenylenediamine reacted similarly to give the analogous benzodiazepines in 70 and 60% yield respectively.

Give a mechanism for the **1** → **2** transformation.

46.   A Furan to Pyran Ring Expansion

The fungal metabolite wortmannin **1** shows a range of interesting and potentially useful biological properties. It has been reported, for example, to be a potent and selective inhibitor of phosphatidylinositol-3'-kinase, which has been identified as an important enzyme in a number of growth factor signalling pathways and is a potential target for intervention in certain proliferative diseases such as cancer. The C-21 position of wortmannin is highly electrophilic, and chemists at Eli Lilly and Company focused on this feature in attempts to modify the furan portion of **1**, which was thought might be responsible for many of its biological activities.

Treatment of wortmannin with trimethylsulfoxonium ylide gave the ring expanded product **2** (R = H). The Lilly chemists considered two possible mechanisms for the formation of **2** and preferred one of them on the basis that use of the perdeuterated sulfoxonium ylide gave exclusively the deuterated product **2** (R = D).

Suggest two mechanisms for the conversion of **1** into **2**, only one of which is consistent with the observed labelling study.

1   **2**

47.   A Structure and Mechanism Correction

It was reported in 1993 that reaction of *N*-(2-bromoethyl)phthalimide with the dianion of isobutyric acid gave the aroylaziridine **1**, and that treatment of **1** with hydrazine in ethanol at 60°C gave the phthalazin-1(2*H*)-one **2**. Tentative mechanisms were suggested for the formation of **1** and **2**. It was subsequently rapidly established, however, that while the condensation of the phthalimide with the dianion was fully reproducible (almost quantitative crude yield), the structure of the product was not **1** as claimed. It was shown that the correct structure for the condensation product was **3**. Moreover, **3** did indeed react with hydrazine to give the phthalazinone **2**.

Give mechanisms for the formation of **3** and the conversion of **3** into **2**. Show also that acceptable mechanisms can be drawn for the formation of **1** and for its conversion into **2**.

1	**2**	**3**

48.   Propargylbenzotriazoles to Five-membered Heterocycles

1-Propargylbenzotriazole, readily available by alkylation of benzotriazole with propargyl bromide in the presence of sodium hydroxide, has been found to be a valuable building block for the synthesis of furans, dihydrofurans, pyrroles and indoles. The versatility of this type of chemistry for the synthesis of otherwise difficultly accessible heterocyclic systems is illustrated by the following typical transformations. A solution of 1-propargylbenzotriazole in THF was treated with one equivalent of BuLi at -78°C for one hour, then one equivalent of 2-bromo-1-phenylpropan-1-one was added and the mixture stirred at -78°C for four hours. A solution of one equivalent of KO^{t}Bu in HO^{t}Bu was added and the temperature was allowed to reach ambient. Reaction was completed by heating the mixture overnight at 50°C. Aqueous work-up gave 2-(benzotriazol-1-ylmethyl)-5-methyl-4-phenylfuran in 53% yield. Stirring of a mixture of one equivalent of this product with one equivalent of zinc chloride and 10 equivalents of 2-methylthiophene in methylene chloride at room temperature followed by aqueous work-up gave 5-methyl-2-(5-methylthiophen-2-ylmethyl)-4-phenylfuran in 86% yield.

Elucidate these reactions and give mechanisms for the formation of the products.

49. From Carvone to an Isoxazoloazepine

Michael addition of cyanide to both (+)- and (-)-carvone **1** has been studied in detail. All four possible products **2** can be obtained depending on the conditions, and it has been shown that **2** (β-CN, β-Me) is the kinetically controlled product, as first described by Lapworth in 1906. Reaction of **2** with pentyl nitrite in the presence of NaOEt/EtOH is strongly exothermic, and the temperature of the reaction mixture must be kept below 0°C. The product, formed in 62% yield, is the isoxazoloazepine **3**.

Give a mechanism for the formation of **3** from **2**.

50. Isoxazoles from Cyclopropanes

Treatment of 2-aryl-1,1-dichlorocyclopropanes **1** under nitrating conditions with mixed acid at 0°C can give 3-aryl-5-chloroisoxazoles in yields of up to 85% depending on the nature of substituents in the aryl ring. The transformation is most efficient when the aromatic substituent is electron withdrawing (R = 4-NO$_2$, 85%; R = 3-NO$_2$, 31%; R = 2-NO$_2$, 13%), but is very poor with electron donating groups (R = 4-Me, 2%; R = 4-MeO, only degradation products).

Suggest a mechanism for isoxazole formation which accounts for these observations.

51.   A Radical Cascade from a Ketene Dithioacetal

Given the selection of an appropriate substrate, the generation of a carbon centred radical from such a substrate can initiate a series of bond-making and bond-breaking processes which are sometimes referred to as radical cascade reactions. These can be of great synthetic value. Thus, treatment of the ketene dithioacetal **1** with a five fold excess of tributyltin hydride in hot benzene under nitrogen and in the presence of a catalytic amount of AIBN gave the metallated benzo[*b*]thiophene **2**, itself a valuable synthetic intermediate, in 70% yield.

Outline the mechanistic pathway for this radical cascade reaction sequence from **1** to **2**.

1   2

52.   A Synthesis of 3-Alkyl-1-naphthols

Standard Heck coupling of *o*-bromopropiophenone with propyne proceeds in 72% yield. Deprotonation of the product thus formed with KHMDS in toluene at -78°C followed by heating at 75°C for one hour gave 3-methyl-1-naphthol in 74% yield after acidic work-up and distillation. This approach has proved to be general and efficient for the preparation of a variety of 3-alkyl-1-naphthols.

Suggest a mechanism for formation of the 3-methyl-1-naphthol which is consistent with the observation that similar cyclisation of the deuterated alkyne **1** gave the naphthol **2**. What conclusion can be drawn from the fact that use of t-butylacetylene instead of propyne in the above sequence gives 3-t-butyl-1-naphthol in 75% yield for the cyclisation step?

1   2

53.   An Entry to Indole Alkaloids of Unusual Structural Type

The *Aspidosperma* alkaloid vincadifformine **1** is reasonably available, and can be readily transformed into the carbinolamine ether **2**. Oxidation of **2** with MCPBA followed by methanolysis gives the hemiketal **3**, and brief treatment of **3** with a 99:1 v/v mixture of CH$_2$Cl$_2$/TFA at room temperature gives a mixture of **4** (42%) and **5** (11%). The yield of **4** is increased to 52%, while almost none of **5** is formed, if the treatment of **3** with acid is allowed to proceed at room temperature for 15 hours. Products **4** and **5** contain the gross skeleton of goniomitine **6**, an indole alkaloid of an unusual structural type.

Give a mechanistic interpretation for the sequence **2** → **3** → **4** + **5**.

1   2   3

4   5   6

54.   Cyclisation as a Key Step to Hirsutene

A key step in the total synthesis of (±)-hirsutene was the acid catalysed (*p*-TsOH/CH$_2$Cl$_2$/RT) cyclisation of **1** to **2**, which proceeded in 95% yield and gave a 10:1 mixture of epimers at C$_4$ (the major epimer was **2**).

Give a mechanism for this transformation which accounts for the observed stereochemistry.

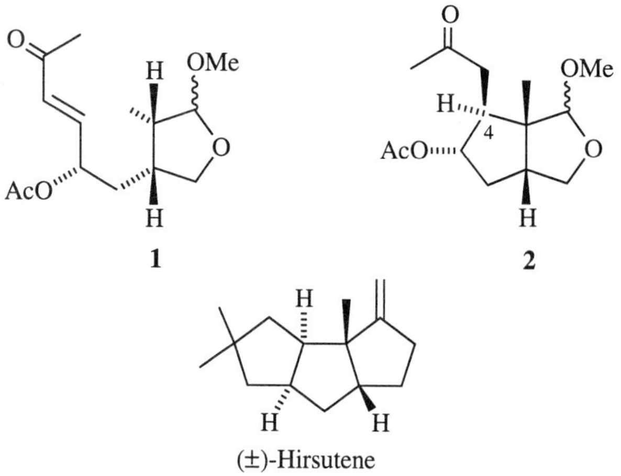

1

2

(±)-Hirsutene

55.   A 1,3-Cyclopentanedione to 1,4-Cyclohexanedione
Transformation

2,2-Dialkylated 1,3-cyclopentanediones undergo facile β-dicarbonyl cleavage on treatment with aqueous alkali, and this process constitutes an excellent synthesis of 5-substituted 4-oxoalkanoic acids. The conversion of **1** into **2** was presumed to proceed in this way for more than 10 years. Closer examination of this transformation, however, revealed the situation to be more complex. When **1** was treated with one equivalent of sodium hydroxide in water at room temperature and the reaction was stopped after 2 minutes, **3** was obtained in *ca.* 50% yield. Treatment of **3** with excess sodium hydroxide solution gave a good yield of **2**.

Give mechanistic explanations for (a) the presumed pathway for conversion of **1** into **2**, and (b) the unusual 1,3-cyclopentanedione → 1,4-cyclohexanedione transformation **1** → **3**.

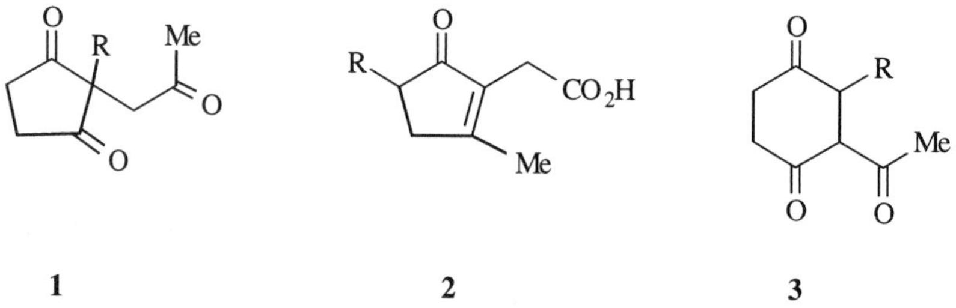

1   2   3

56.   Conversion of *o*-Hydroxyaryl Ketones into 1,2-Diacylbenzenes

Treatment of acylhydrazones of *o*-hydroxyaryl ketones with LTA in THF at room temperature results in smooth, very high yield conversion into 1,2-diacylbenzenes, e.g. **1** → **2**. The mechanism of this very unusual replacement of a phenolic OH group by an acyl group has been studied in detail, and standard crossover experiments using, for example, a mixture of **1** and its 3,5-dibromo derivative established that the process was intramolecular. Use of **1** in which the oxygen atom of the benzoyl group was labelled as ^{18}O gave exclusively the labelled product **3**.

Suggest a mechanism for the **1** → **2** conversion which is consistent with these observations.

57.   Mechanisms of Bimane Formation

"Bimane" is the trivial name given to the 1,5-diazabicyclo[3.3.0]octadienediones **1**. The first examples of these compounds were prepared almost a century ago and they have attracted considerable attention in recent years because they are not only very stable but the *syn*-isomers show "a beautiful and striking fluorescence" whereas the *anti*-isomers are only weakly fluorescent. The first general synthesis of bimanes was described in 1978, when it was shown that treatment of a methylene chloride solution of 4-chloropyrazolin-5-ones with two equivalents of $K_2CO_3.^1/_2H_2O$ and 0.75 equivalent of K_2CO_3 at 0°C for 18 hours gave good yields of the *syn*-bimanes **1** together with small amounts of the *anti*-isomers.

In 1996, the tetraphenyl *syn*-bimane **1** (R = Ph) was obtained in 19% yield when the sodium salt of the 1,3,4-oxadiazinone **2** was treated with di-t-butyl dicarbonate in THF in an attempt to form the 3-Boc derivative. A mechanism was suggested for the conversion of **2** into *syn*-**1** (R = Ph) which involved - ultimately - an intermediate of similar type to that which had been suggested previously for the transformation of 4-chloropyrazolin-5-ones into **1**.

Suggest mechanisms for both types of transformation.

syn-**1**   anti-**1**   **2**

58.　A Carbohydrate to Cyclopentanol Conversion

There is much current interest in the development of simple and effective methods for the conversion of selectively functionalised carbohydrates into stereodefined cyclopentanols, and South African chemists have reported an excellent samarium diiodide-based process. Thus, dropwise addition of the iodoglucoside **1** to a refluxing solution of excess of samarium diiodide in THF/HMPA, and heating of the mixture under reflux for 2 hours followed by cooling, dilution with 1:1 hexane/ethyl acetate and quenching with 5% aqueous citric acid gave the pure cyclopentanol **2** in 70% yield after flash chromatography.

Suggest a mechanism for the transformation of **1** into **2**, which has been described as a "SmI$_2$-promoted Grob-fragmentation reaction . . . . followed by an *in situ*, stereocontrolled SmI$_2$-mediated cyclisation . . . . "

1   **2**

59.   Pyridazines from 1,2,4-Triazines

Treatment of 3-chloro-6-phenyl-1,2,4-triazine **A** with 1.1 equivalents of phenylacetonitrile in dry *N,N*-dimethylacetamide at 0°C in the presence of an excess of KO^{t}Bu for one hour followed by quenching of the reaction mixture with ice-water gave an 86% yield of a dinitrile **B**, C$_{17}$H$_{12}$N$_4$. Treatment of **B** with 1:1 aqueous ammonia/acetone for one hour resulted in quantitative conversion into 3-amino-4,6-diphenylpyridazine, **C**. The triazine **A** was converted directly into the pyridazine **C** when DMF was used as solvent and the reaction mixture was quenched with aqueous acetic acid.

Identify the intermediate **B** in the dimethylacetamide reaction, and suggest a mechanism for the overall conversion **A** → **B** → **C**.

60.   Natural Product Degradation During Extraction and/or Chromatography

The shrub *Baeckea frutescens* (Myrtaceae) has long been used in traditional medicine for the treatment of rheumatism and snake bites, and a recent examination of a dichloromethane extract of the aerial parts revealed the presence of a number of structurally unique compounds. The major component of the extract was shown to have structure **1**. Another constituent of the mixture was shown to have structure **2**, and it was tentatively suggested that **2** might be a degradation product of **1**, formed either during the extraction process or during chromatographic separation of the crude plant extract.

Suggest a possible mechanism for the conversion of **1** into **2**.

1   2

61.   Simple Access to Oxaadamantanes

Oxaadamantane and its derivatives are molecules of considerable interest to chemists and biologists, but until recently there was no convenient route to such compounds. Such a route has now been discovered, starting from the readily accessible 2-hydroxy-2-methyladamantane **1**. Thus, oxidation of **1** by trifluoroperacetic acid in TFA at 20°C for one hour gave a mixture (actual yield not given; merely described as "a good yield") of three products : oxaadamantane **2** (56%), *exo*-4-hydroxy-2-oxaadamantane **3** (40%) and the lactone **4** (4%).

Suggest mechanisms to account for the formation of **2**, **3** and **4**. Note that oxaadamantane **2** is stable under the reaction conditions, and does not serve as a precursor to the hydroxy derivative **3**.

62.   A Pyrimidine Rearrangement

6-Substituted uracils **1** are of much interest due to their possible use as anti-cancer and anti-AIDS drugs, and both 2,4-dimethoxy-6-iodopyrimidine and 1,3-dimethyl-6-iodouracil (**1**, R = Me, X = I) were required as starting materials for the synthesis of a variety of uracils **1** by palladium-catalysed C-C bond formation. It was reported in 1961 that treatment of 6-chloro-2,4-dimethoxypyrimidine with sodium iodide in refluxing DMF gave a 42% yield of the corresponding 2,4-dimethoxy-6-iodopyrimidine, but repetition of the reaction recently clearly established that the product was in fact the isomeric 1,3-dimethyl-6-iodouracil (**1**, R = Me, X = I).

Give a mechanism to account for this transformation.

1

63. 1,4-Thiazin-3-ones from 1,4-Oxathiins

1,4-Thiazin-3-ones **2** show "a remarkable spectrum of biological activity", but there are few convenient methods for their preparation. A simple and highly effective method has now been described which involves brief heating of 1,4-oxathiin carboxamides **1** with concentrated hydrochloric acid in acetonitrile at 80°C. Yields of the thiazin-3-ones **2** are generally excellent.

Suggest a mechanism for the **1 → 2** transformation.

1   **2**

64. An Unusual Route to 3-Acylfurans

Simple 3-acylfurans are not accessible by direct Friedel-Crafts-type substitution of furan, but a most unusual route to these compounds has been discovered. Thus, reaction of the stannylated cyclopropane derivative **1** with acid chlorides (RCOCl) in refluxing toluene for 5 hours gave the dihydrofurans **2** in 64-83% yield. Treatment of **2** with 1.2 equivalents of BF₃.OEt₂ in dichloromethane at -78°C followed by slow warming of the reaction mixture to room temperature gave the 3-acylfurans **3** in 30-62% yield.

Give a mechanistic interpretation for the sequence **1 → 2 → 3.**

1   **2 (R = alkyl, aryl)**   **3**

65. Reaction of PQQ with L-Tryptophan

Pyrroloquinolinequinone (PQQ, **1**) is one of a number of *o*-quinones which serve as prosthetic groups in quinoproteins, and has been the subject of intense interest because of its growth-stimulating, pharmaceutical and nutritional activities (c.f. Problem number 41). There is also considerable interest in the *in vitro* reactions of PQQ with biomolecules, and in this context the reaction of PQQ with L-tryptophan in phosphate buffer (pH 6.5) under aerobic conditions was investigated. This gave three products **2**, **3** and **4**, with **2** the major component. Compounds **2** and **4** showed a much more pronounced growth-stimulating effect than PQQ, while **3** also showed "a marked effect".

Give mechanisms for the reaction of **1** with L-tryptophan to produce **2**, **3** and **4**.

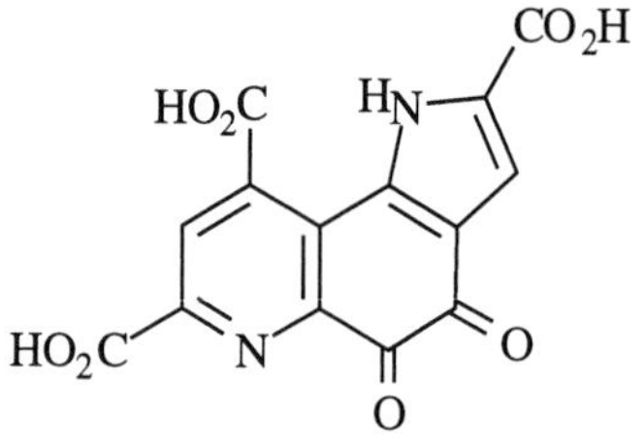

1

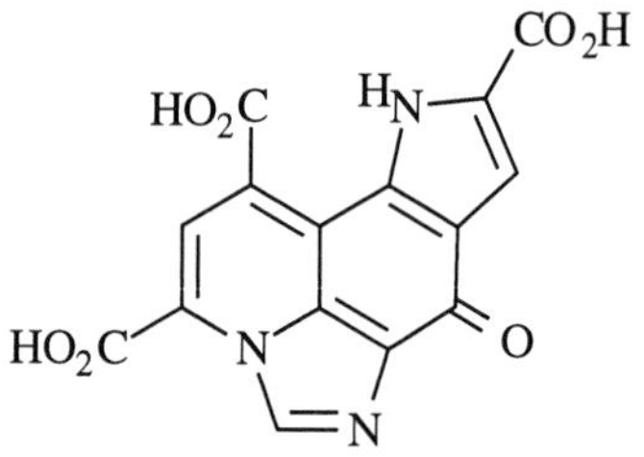

2

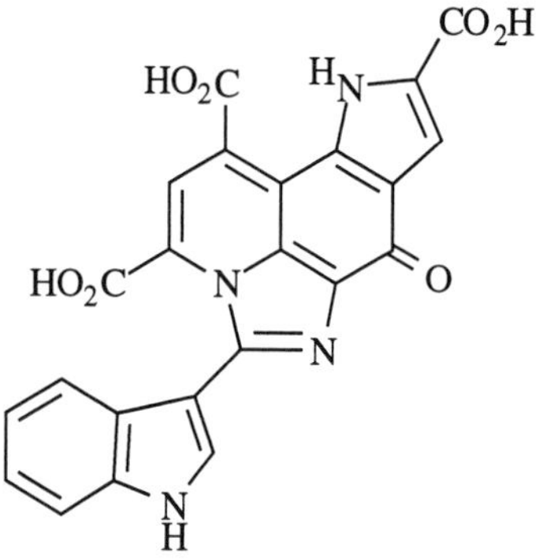

3

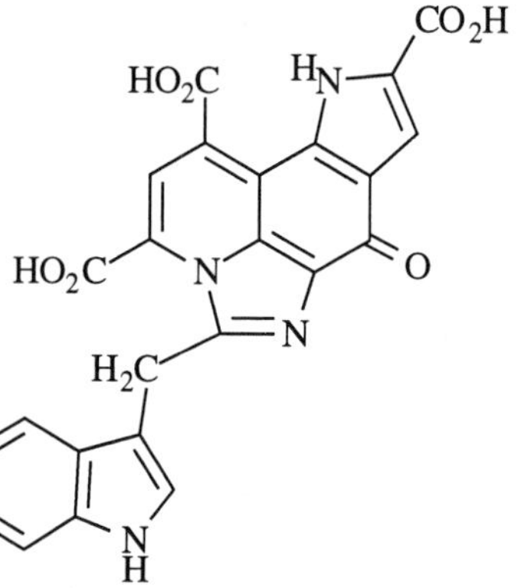

4

66.   An Unexpected Result from an Attempted Double Bischler-Napieralski Reaction

As part of a programme designed to produce compounds with axial chirality, double Bischler-Napieralski reactions were carried out with oxamide derivatives of ω-arylalkylamines. Thus, treatment of the oxamide derived from 3,4-dimethoxy-β-phenylethylamine with pyrophosphoryl chloride in acetonitrile gave **1** in 84% yield, as expected. An attempt was then made to extend this double cyclisation protocol to the oxamide derived from 2-(3-methoxyphenoxy)ethylamine. Reaction of this latter compound under the same conditions used for the formation of **1**, however, gave **2** in 81% yield instead of the expected product of a double Bischler-Napieralski reaction.

Suggest a mechanistic explanation for the formation of **2**.

1

2

67.   Bis-Indole Alkaloid Formation

The biosynthesis of many bis-indole alkaloids has been postulated to proceed by dimerisation of appropriate precursors, and there is now a substantial amount of experimental evidence to support this hypothesis. For example, treatment of the alcohol **1** with acid gives the alkaloid yuehchukene **2**, and **1** could arise biogenetically by *in vivo* prenylation of indole followed by enzymatic oxidation. A study of related 2-prenylated indoles has confirmed the ease with which such molecules can "dimerise". Thus, treatment of the secondary alcohol **3** in benzene with silica gel impregnated with TsOH gave a complex mixture of products from which **4** (5.1%) and **5** (2.1%) were isolated (**3** is very sensitive to acid, and is easily decomposed). Treatment of the isomeric tertiary alcohol **6** with a catalytic amount of TFA in anhydrous benzene gave much higher yields of the two "dimeric" products **7** (31%) and **8** (25%).

Give mechanistic explanations for the conversion of the alcohol **1** into the alkaloid yuehchukene **2**, and for the acid catalysed transformations of **3** into **4** and **5** and of **6** into **7** and **8**.

1   **2**   **3**

4   **5**   **6**

7   **8**

68.   Azaphenalene Alkaloid Synthesis : A Key Step

Precoccinelline **1** is one of a number of azaphenalene alkaloids which are responsible for the bitter taste of many coccinellids, and may be implicated in the defence mechanism of the beetles. Coccinellids are apparently the only source known so far of alkaloids based on the azaphenalene skeleton. A key step in one very elegant synthesis of precoccinelline **1** involved reaction of the bicyclic system **2** with TBDMSOTf and 3-[2-(1,3-dioxolanylpropyl)]magnesium chloride at -78°C in a mixed $CH_2Cl_2/Et_2O/THF$ solvent system followed by quenching with saturated ammonium chloride solution. This gave **3** as the major product in 74% yield.

Suggest a mechanism for the transformation of **2** into **3**.

$$R = (CH_2)_3 - \begin{array}{c} O \\ \diagup \\ O \end{array}$$

69.   A Primary Nitroalkane to Carboxylic Acid Transformation

The Nef oxidation of nitroalkanes is a valuable method for the preparation of aldehydes and ketones. French workers have now shown that primary nitroalkanes, used as such or generated *in situ* from the corresponding bromides, can be oxidised directly to carboxylic acids. Yields are generally excellent and the conditions are very mild; acids, esters, alkenes, a TMS-protected alkyne, alcohols, THP ethers, and an imide were unaffected. Thus, stirring of a mixture of 2-nitroethylbenzene (1 mmol) and sodium nitrite (3 mmol) in acetic acid (10 mmol) and DMSO (2 ml) for 6 hours at 35°C followed by acidification with 10% aqueous hydrochloric acid gave phenylacetic acid in 95% yield. Use of 2-bromoethylbenzene in place of 2-nitroethylbenzene gave the same product in 85% yield.

Give a mechanism to account for this oxidation.

70.   α-Diazoketones with Rhodium(II) Acetate

There has been great interest in recent years in methods for the generation of azomethine ylides and in exploitation of these reactive species in tandem/cascade processes for the rapid assembly of polyaza, polycyclic, multifunctional systems. α-Diazo ketones have featured greatly in such studies, treatment with a catalytic amount of rhodium(II) acetate generating transient rhodium carbenoids. A very common feature of many investigations of this type is the occurrence of quite unexpected reactions. For example, treatment of the diazo ketone **1** with a catalytic amount of rhodium(II) acetate in chloroform in the presence of DMAD gave, unexpectedly, the three products **2**, **3** and **4** in 15, 18 and 60% yield respectively.

Deduce the structure of the "expected" product from this reaction and suggest mechanisms for the conversion of **1** into **2**, **3** and **4**.

1

2

3

4

71. Pyrrolo[1,2-*a*]benzimidazole Synthesis by Ortho Nitro Interaction

Pyrrolo[1,2-*a*]benzimidazoles are of interest as potential antitumor agents, and one convenient and short route to such compounds involves treatment of the readily available 1-(*o*-nitroaryl)pyrrolidines **1** with ZnCl$_2$/Ac$_2$O at 110°C for several hours. This gives the 3-acetoxy derivatives **2**, usually in modest yield, and the initial cyclisation reaction is known to involve a ZnCl$_2$-catalysed internal redox reaction leading to an *o*-nitroso iminium ion. Further study of this process has allowed a more complete mechanism to be postulated which can account for formation of the 3-acetoxy derivatives. Thus, use of **3** as substrate in the ZnCl$_2$/Ac$_2$O reaction gave the three products **4**, **5** and **6** in 16, 14 and 9% yield respectively.

Suggest mechanisms which account for these observations.

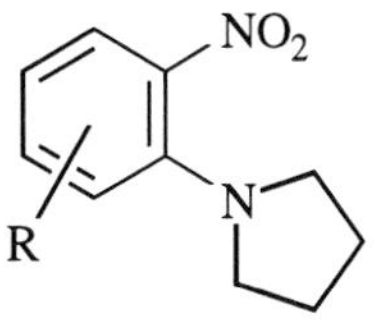

1 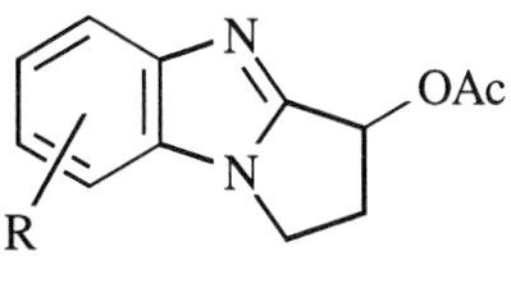   2 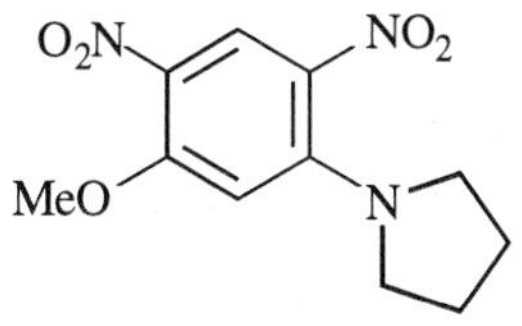   3

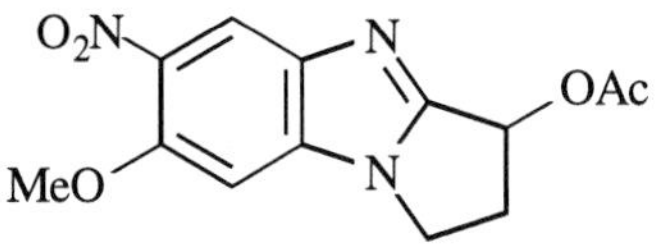

4 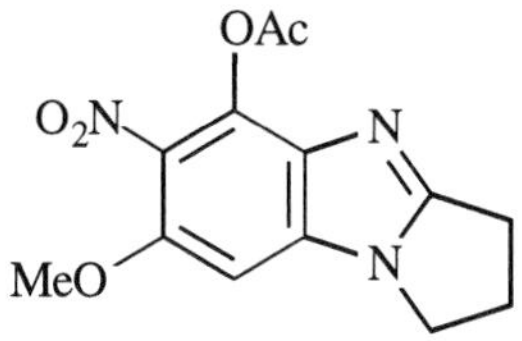   5

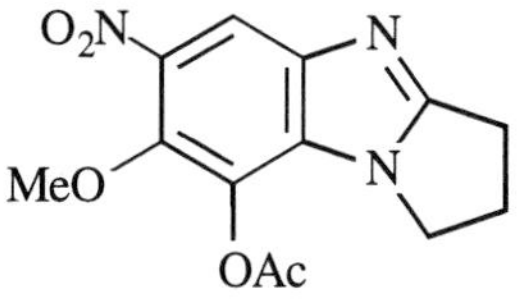

6

72.   Horner-Wittig Reaction on a Bifuranylidenedione

Reaction of the diarylbifuranylidenedione **1** with two equivalents of the stable ylide Ph$_3$PCHCO$_2$Me in refluxing toluene under nitrogen for 24 hours gave a mixture of products, among which were the expected mono-Horner-Wittig ester **2** (26%) and the tricyclic product **3** (11%).

Suggest a mechanism for the formation of **3**.

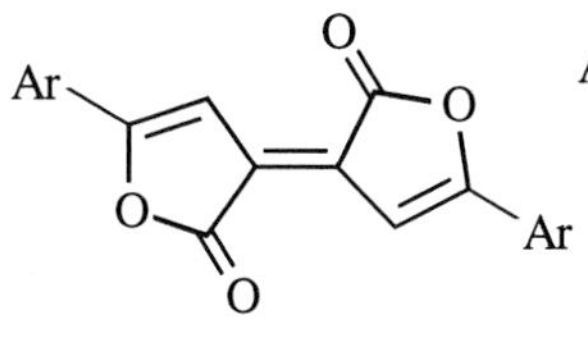

1 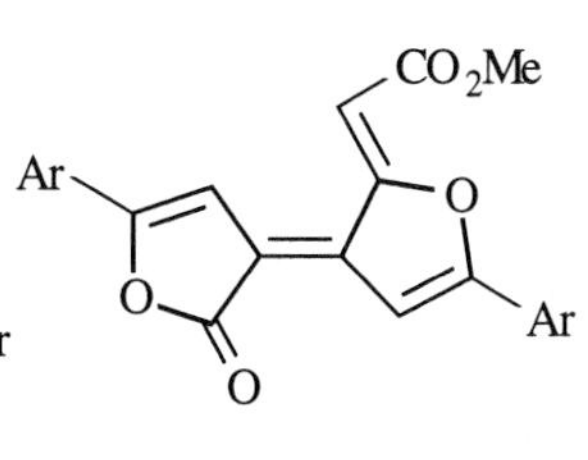   2 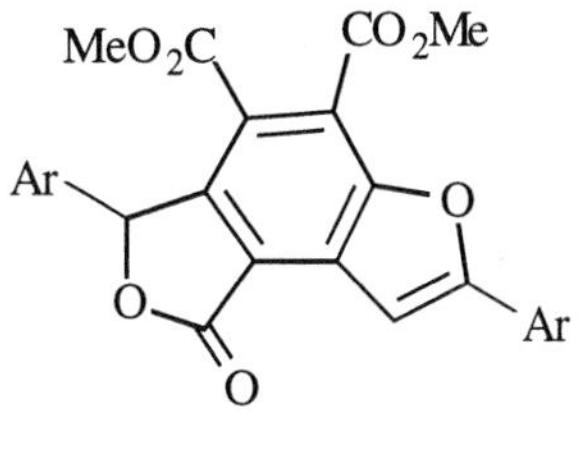   3

Ar = mesityl

73.   Lactone Ammonolysis

The lactone **1** was subjected to ammonolysis by treatment at room temperature for 5 hours with ethanol saturated with ammonia in the expectation that simple ring opening would occur. The product was not, however, the expected amide; instead, two products were formed, the amide **2** and (*R*)-lactamide, in almost quantitative yield. An unstable intermediate could be isolated and was shown to serve as precursor to **2**.

Suggest a mechanism for the formation of **2** from **1**.

1   **2**

R = PhCH₂O

74.   Intramolecular Schmidt Reaction

In a one-pot process which has been described as an intramolecular Schmidt reaction it has been shown that treatment of ketals or enol ethers of 1,5-azidoketones with Lewis or protic acids followed by sodium iodide in acetone results in formation of lactams in 68-95% yield.

Give a mechanism for this reaction, a representative example of which is shown.

75.   Indolizidones by Intermolecular Photochemical Reaction

Anthraquinone photosensitised irradiation of tertiary *N*-allylamines in the presence of αβ-unsaturated esters gives good yields of lactams. Using pyrex-filtered output (λ>290 nm) from a 450 W medium pressure Hanovia lamp, for example, *N*-allylpiperidine and methyl methacrylate reacted in acetonitrile to give a 2:3 diastereomeric mixture (2*S*, 9*S* : 2*R*, 9*S*) of the indolizidone **1**.

Suggest a mechanism to account for the formation of **1**.

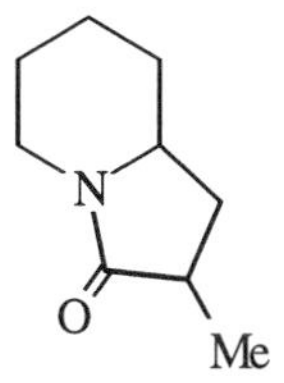

1

76.   An Unexpected Product from Amine Oxidation

Treatment of *N*-methylaniline with 4-chlorobut-2-yn-1-ol in acetone at reflux temperature in the presence of anhydrous K_2CO_3 gave the expected tertiary amine in 91% yield. Room temperature oxidation of a dilute solution of this amine with MCPBA resulted in formation of a colourless crystalline solid in 56% yield, which was shown to have structure **1**.

Deduce a mechanism for the formation of **1**.

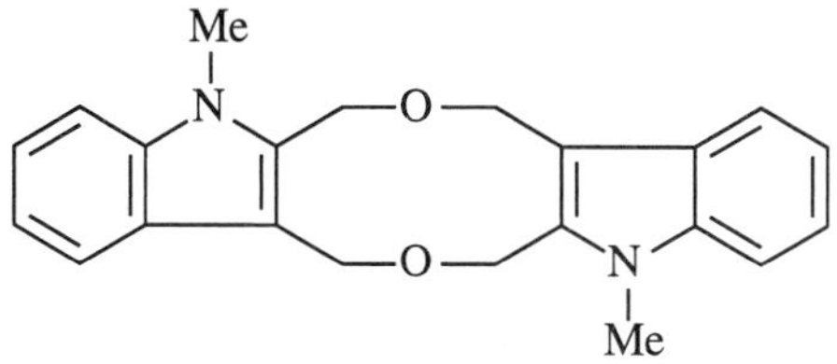

1

77.   Degradation of Pyripyropene A

Pyripyropene A, **1**, is one of a group of related natural products, the pyripyropenes, which are the most potent known natural product-derived inhibitors of acyl-CoA : cholesterol acyltransferase. Biosynthetic studies using ^{13}C and ^{14}C labelled precursors established that **1** is derived from three mevalonates, five acetates and one nicotinic acid, and a degradation method described as "unique" was used to reisolate the intact labelled nicotinic acid precursor from **1** produced in a culture broth of *Aspergillus fumigatus*. Thus, **1** produced from [carboxy-^{14}C]nicotinic acid was oxidised to the 13-oxo derivative with Jones's reagent, and the ketone was treated with sodium methoxide. After appropriate work-up and purification, [carboxy-^{14}C]-nicotinic acid containing all of the original label was isolated together with the inactive pyrone **2**.

1	**2**

Give a mechanistic explanation for these observations.

78.   Benzofurans from Cyclobutenediones

Complex polyfunctional molecules can often be assembled efficiently by short, spectacular sequences of reactions, an example of which is the preparation of the pentasubstituted benzofuran **1**. Thus, addition of 1-lithio-1-methoxy-3-(trimethylsilyl)-1,2-hexadiene to 3,4-dimethoxycyclobut-3-ene-1,2-dione gave the expected keto alcohol in 70% yield. This alcohol was heated at reflux temperature in toluene for 4 hours to give a 2,3,5,6-tetrasubstituted hydroquinone in 90% yield. Oxidation of the hydroquinone with silver oxide and potassium carbonate in anhydrous benzene (90%) followed by reaction of the quinone thus obtained with TFA in methylene chloride at 0°C then at room temperature for two days gave **1** in 75% yield.

Elucidate the overall synthetic scheme and give mechanisms for each of the steps.

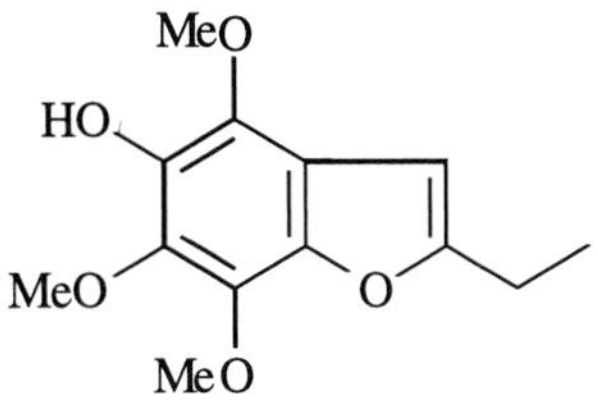

1

79. Acid-Catalysed Condensation of Indole with Acetone

The hydrochloric acid-catalysed condensation of indole with acetone has been a topic of occasional study for almost a century. A complex mixture of products is obtained, but using modern separation, analytical and spectroscopic techniques it is now possible to assign structures unambiguously. The latest compound to be identified is **1**.

Suggest a mechanism to account for the formation of **1** from indole and acetone.

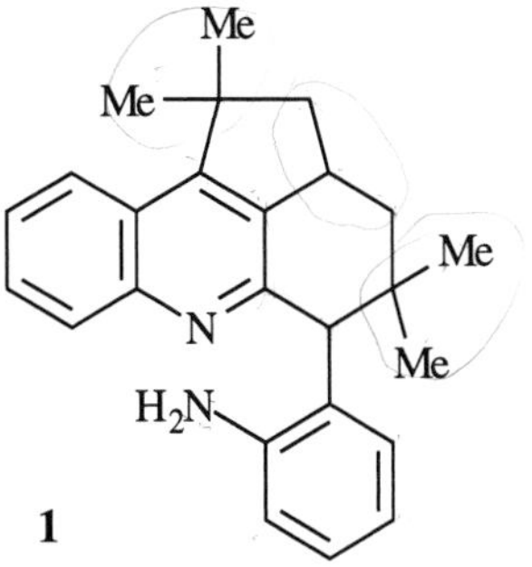

1

80. A Synthesis of 2-Formylpyrrolidines

Reaction of the imines **1** with NCS in CCl_4 resulted in rapid monochlorination in almost quantitative yield. Treatment of the dichlorides thus obtained with 1.1 equivalent of K_2CO_3 in methanol under reflux for 3.5 hours smoothly produced the pyrrolidines **2**, from which the 2-formyl derivatives could be obtained by acid hydrolysis.

Give mechanisms for the overall conversion of **1** into **2**.

$$Cl(CH_2)_3CHCH = NR^1$$

with R^2 substituent

and

$$N(R^1) \text{ ring with } R^2 \text{ and } CH(OMe)_2$$

1   **2** $R^1, R^2 = alkyl$

81.   An Abnormal Claisen Rearrangement

As part of a medicinal chemistry programme to synthesise analogues of mycophenolic acid **1**, the allyl ethers **2a-c** were subjected to Claisen rearrangement and gave the expected products in moderate to good yield. When the allyl ether **2d** was subjected to the same conditions, however, the only rearranged product which was obtained was the phthalide **3** (PhNEt$_2$ at 200°C, 35% yield). Addition of an equimolar amount of HMDS to the mixture raised the yield of **3** to 81%. Re-examination of the reactions of **2a** and **2b** showed that small amounts (9-16%) of rearranged products corresponding to **3** could be produced in these reactions as well as the major products of normal Claisen rearrangement.

Suggest a mechanism for the conversion of **2d** into **3** and account for the increase in yield which is observed when HMDS is added to the reaction mixture.

1

2

a, R - allyl
b, R = *cis*-2-butenyl
c, R = 3-cyclohexenyl
d, R = 3-methyl-2-butenyl

3

82.   Benzoxepinones from Phthalides

Addition of an ethereal solution of prop-2-ynylmagnesium bromide to phthalides **1** at 0°C followed by quenching of the reaction mixture with 20% HCl results in direct formation of the benzoxepinones **2** in 45-86% yield.

Give a mechanistic explanation for this transformation.

R = H, alkyl, vinyl

83.   *m*-Terphenyls from Pyrones

Alkali-catalysed reaction of the pyrones **1** with acetophenones Ar^1COMe in DMF at room temperature results in smooth transformation into the *m*-terphenyls **2** (50-66%) and the pyranylideneacetates **3** (0-28%).

Give mechanisms to account for the formation of **2** and **3** from **1**.

84.   A Cascade Reaction

Treatment of the readily accessible ester **1** with either Ac_2O/TsOH/refluxing xylene or TFAA/Et_3N/CH_2Cl_2/RT or TMSOTf/Et_3N/CH_2Cl_2/RT results in smooth conversion into the indoline **2** in 50-75% yield depending on the conditions employed.

Explain this transformation in mechanistic terms.

            1   **2**

85.   Photochemical Ortho Rearrangements

Irradiation of *o*-acetylphenylacetonitrile at λ>280 nm in methanol containing "a few percent water" followed by chromatography over silica gel gave *o*-acetylphenylacetamide in more than 80% yield. When the reaction was repeated with methanol containing only 0.02% water, and the reaction mixture was worked up without use of acid or silica gel chromatography, the major product was the dimethyl ketal of *o*-acetylphenylacetamide. In a further experiment, irradiation of ^{18}O-labelled *o*-acetylphenylacetonitrile gave a mixture of *o*-acetylphenylacetamide and its dimethyl ketal in which all of ^{18}O label was located in the amide carbonyl oxygen.

Give a mechanistic interpretation of these reactions.

86.   From Bicycle to Pentacycle to Tricycle

Reaction of the lactol **1** with the isocyanate **2** in acetonitrile in the presence of 5 mol % of DBU resulted in a cascade of reactions and formation of a pentacyclic product **A** in 86% yield. Treatment of **A** with methoxide gave the highly functionalised octahydroacridine **3** in 90% yield.

Identify the pentacycle **A** and give mechanisms for (a) its formation, and (b) its conversion into **3**.

            1   **2**   **3**

87. Lewis Acid-Catalysed Cyclopropyl Ketone Rearrangement

In studies directed towards potential routes to the 11-oxosteroid framework it was found that treatment of the ketene dithioacetal **1** with tin(IV) chloride in benzene at 20°C followed by aqueous quench led directly to **2** in 82% yield.

Suggest a mechanism for the **1** → **2** conversion.

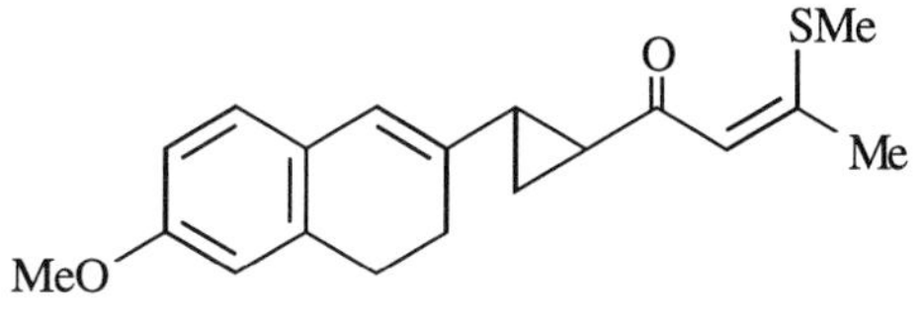

1

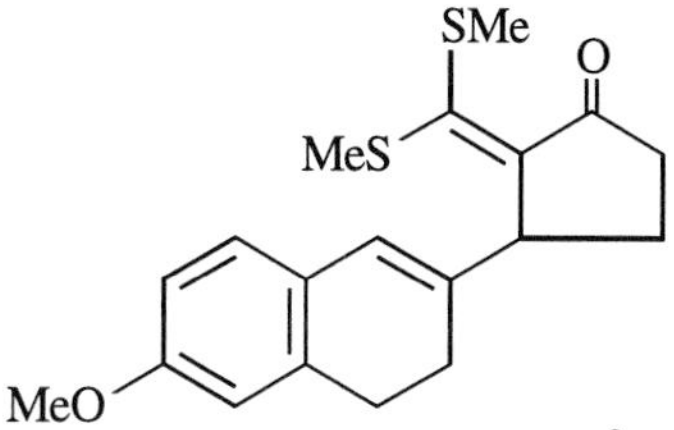

2

88. An Intramolecular Ring Closure

Treatment of the amide **1** with TFAA in methylene chloride at 25°C gave **2** in 85% yield.

Suggest a mechanism for this transformation.

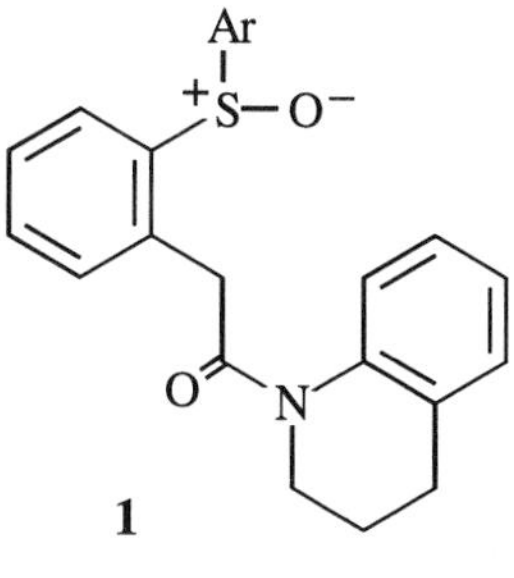

1

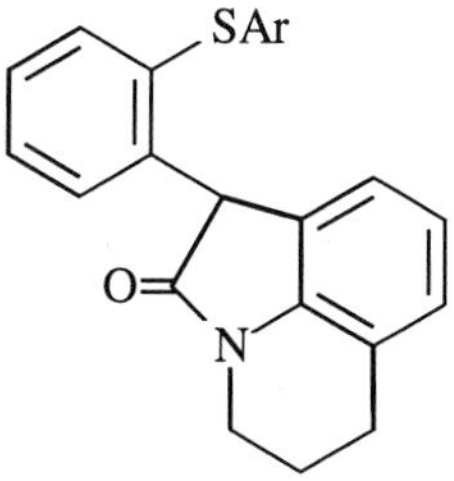

2

Ar = 4-MeC$_6$H$_4$

89. Photochemical Rearrangement of Pyran-2-ones

The photochemical behaviour of pyran-2-ones continues to attract interest and to yield unusual types of reaction. In a recent example it was found that irradiation of the 6-(2-hydroxyalkyl)pyran-2-ones **1** in methanol under carefully optimised conditions (which were not clearly specified) gave a diastereomeric mixture of solvent adducts. Removal of the solvent and treatment of the residue with a catalytic amount of HCl in THF at room temperature resulted in formation of the dihydropyrans **2** in good yield.

Suggest a mechanism to account for this transformation.

1

2

R = alkyl, aryl

90.   A Paclitaxel Rearrangement

There has been intense interest in recent years in selective modification of functionality in baccatin III, **1**, the basic diterpenoid core of paclitaxel (Taxol®) which is now established as a clinically active antitumour drug. Given the complexity of **1**, rearrangements are common even under mild conditions and attempts to carry out apparently simple transformations are frequently frustrated, as shown by the following example. The 13β-chloro derivative **2** was prepared and fully characterised, then treated with sodium azide in aqueous DMF at 60°C in the expectation that the 13α azide would be obtained. The product, however, was shown to be the ring cleaved compound **3** (71% yield).

Suggest a mechanism to explain this transformation.

1, $R^1 = R^3 = H$; $R^2 = OH$
2, $R^1 = Cl$; $R^2 = H$; $R^3 = SiEt_3$

3

91. An Ionic Diels-Alder Reaction

In an interesting and potentially very useful extension of the so-called ionic Diels-Alder reaction it was shown that reaction of acetals of the type **1** with various 1,3-dienes **2** in the presence of an acid catalyst followed by hydrolysis with TsOH in methanol gave good yields of cycloadducts **3**. In most cases HBF$_4$.OEt$_2$ was found to be the most efficient catalyst and diastereomeric ratios as high as 200:1 could be achieved.

Give a mechanistic explanation for this ionic Diels-Alder reaction.

1. Acid catalyst
2. H$^+$/MeOH

1   **2**   **3**

R = ^{i}Pr, ^{t}Bu

92. The Eschenmoser Fragmentation Reaction Extended

A simple, high yielding (65-90%) procedure has been described for the preparation of medium ring and macrocyclic acetylenic lactones which is mechanistically related to the well known Eschenmoser fragmentation reaction, and the following example is representative. Bromination of the tosylhydrazone of **1** was carried out with NBS at -10°C in a water/t-butanol/acetone mixture. The reaction mixture was then treated with aqueous NaHSO$_3$ solution and the resulting mixture heated at 50-60°C for one hour, which gave the acetylenic lactone **2**.

Give a mechanism for this transformation.

1   **2**

93.   A Modified Batcho-Leimgruber Synthesis

The Batcho-Leimgruber indole synthesis is probably the most important and widely used method for the preparation of 1*H*-2,3-unsubstituted indoles. In an attempt to extend the utility of the process to the preparation of 3-substituted indoles, functionalisation of the specific intermediate enamine **1** was explored. Treatment of **1** with ethyl chloroformate in refluxing chloroform in the presence of *N,N*-diethylaniline failed to give the expected ester. Instead, a product was obtained the spectral data for which suggested that it was a dienamine. The same product could be obtained in 52% yield simply by treatment of **1** with half an equivalent of TsOH in toluene at 90°C, and catalytic hydrogenation of this dienamine gave **2** in 58% yield.

Identify the dienamine and give a mechanism for its formation.

1   2

94.   A Pyrone to Pyran Conversion

The highly functionalised pyran **1** can be prepared in a one-pot operation and in yields up to 30% by reaction of the readily available sulfone **2** with methyl coumalate **3**, using methylene chloride as solvent and DBU as base.

Give a mechanistic explanation for the formation of **1**.

MeCOCH₂SO₂Ph

1   2   3

95.   Conformationally Rigid 4-Oxoquinolines

A number of 4-oxoquinoline-3-carboxylic acids are commercially very important antibacterials and there is continuous research to extend the range of clinically useful products. As part of a programme on the synthesis of conformationally rigid analogues of the basic system it was found that heating of the amide **1** in acetic anhydride at 80°C for 4 hours gave a 92% yield of the tetracyclic analogue **2**.

Suggest a mechanism for the conversion of **1** into **2**.

96.   An Efficient Chromene Synthesis

Lithiation of commercially available phenyl vinyl sulfoxide with LDA followed by reaction with sodium 2-acetyl-4-methylphenoxide gave the expected alcohol in 90% yield. This was not very stable but was nevertheless heated under reflux in ethanolic sodium ethoxide (excess) for 19 hours. Standard work-up and purification gave the chromene **1**, also in 90% yield.

Give a mechanism for the formation of **1**.

97. Flash Vacuum Pyrolysis of *o*-Xylylene Dimers

Flash vacuum pyrolysis of [4+4] dimers of *o*-xylylenes gives anthracenes as the major products, and this "remarkable transformation" is highly regiospecific. Thus, the methyl-substituted derivative **1** gives **2** while **3** gives **4**.

Account for these observations in terms of mechanism.

1	**2**

3	**4**

98. Synthesis of a Neocarzinostatin Building Block

The naphthoic acid **1** is a key building block required for the synthesis of the chromophore component of the enediyne antitumor agent neocarzinostatin, but has until recently only been available by long and not very efficient syntheses. The following route can be used to prepare tens of grams of **1** : room temperature, CDI-catalysed condensation of magnesium methyl malonate with 3-(2-methyl-4-methoxyphenyl)prop-2-enoic acid gave the expected keto ester (78%) which was converted into the monochloro derivative (92%) by treatment with sulfuryl chloride in benzene at 66°C. Irradiation of this latter intermediate for one hour at room temperature with a 450-watt Hanovia medium pressure arc lamp gave the methyl ester of **1** in 53-63% yield.

Elucidate the overall synthesis and give a mechanism for the final step.

1

99.   Oxazole Esters from α-Amino Acids

The following standard esterification procedure was applied to *N*-benzoylalanine: oxalyl chloride was added to a solution of *N*-benzoylalanine in anhydrous THF, the mixture was stirred overnight at room temperature, then the solvent and residual traces of oxalyl chloride were evaporated *in vacuo*. Triethylamine and then methanol were added to the residue at 0°C. On work-up, none of the expected amino acid ester was obtained. The product, isolated in 51% yield, was found to be methyl 2-phenyl-4-methyloxazole-5-carboxylate, and the process was found to be general for a variety of *N*-benzoylated α-amino acids.

Suggest a mechanism for this transformation.

100.   Unexpected Formation of a Phenazine

During a study designed to investigate whether reaction of the weakly nucleophilic diisopropylamido anion with sulfenamides might proceed by a single electron transfer process and result in formation of aminyl radicals, it was found that addition of 4'-methoxybenzenesulfenanilide, 4-MeOC$_6$H$_4$NHSPh to a solution of 1.5 equivalents of LDA in THF at -20°C under nitrogen followed by stirring at room temperature overnight gave a homogeneous, colourless solution. This was then exposed to air, whereupon the solution immediately turned deep red in colour. The mixture was stirred in air for 2 hours at room temperature then washed with brine and extracted with ether. Chromatography of the crude product gave 2,7-dimethoxyphenazine in 95% yield.

Suggest a mechanism to account for this transformation.

101.   Conversion of Primary Amides into Nitriles

Many methods for the dehydration of primary amides to nitriles involve the use of strong acids and/or other reagents which are incompatible with sensitive functionality. French workers have recently described an unusual, mild, efficient (64-92%) and fairly general procedure for the RCONH$_2$ → RCN transformation which simply involves heating a mixture of the amide, an aldehyde and formic acid under reflux in acetonitrile for 12 hours. The following observations were made:

(a) Benzonitrile can be used as solvent in place of acetonitrile. Benzamide is formed in an amount equivalent to that of the primary amide.

(b) The nature of the aldehyde used is not critical: 1-octanal is as effective as paraformaldehyde.

(c) The aldehyde component is required in only catalytic amounts.

(d) Formic acid appears to be essential, although the reason is not known. Acetic and trifluoroacetic acids are ineffective.

Give a mechanistic interpretation for this $RCONH_2 \rightarrow RCN$ conversion.

102. An Efficient Benzo[*b*]fluorene Synthesis

Heating of a mixture of 2-phenyl-1,4-naphthoquinone (1 eq.) with dimethyl malonate (4 eq.) and manganese(III) acetate (6 eq.) in acetic acid at 80°C for 16 hours gives **1** in 76% yield, and this has been shown to be a general type of reaction for a variety of 2-aryl-1,4-naphthoquinones.

Suggest a mechanism for the formation of **1**

1, R = CO$_2$Me

103. A Facile Synthesis of Tetracyclic Pyrroloquinazolines

Condensation of 2-aminobenzylamine with methyl 3,3,3-trifluoropyruvate gave a mono-adduct $C_{11}H_{13}F_3N_2O_2$ in almost quantitative yield which reacted smoothly with cyclohexanone in diethyl ether at room temperature to give a mixture of **1** and **2** in 38 and 28% yield respectively.

Suggest a mechanism for the formation of **1** and **2**.

| **1** | **2** |

104. Ethyl 2-Chloronicotinate from Acyclic Precursors

Most pyridine derivatives used on a commercial scale are prepared by functionalisation of available pyridine compound feedstocks. This can often be problematic, not least in terms of regioselectivity in the introduction of substituents. Attention is therefore increasing on methods for the efficient construction of pyridine derivatives from cheap acyclic precursors, and a recent case concerns 2-chloronicotinic acid, a key building block for many agricultural and pharmaceutical products. The ethyl ester of 2-chloronicotinic acid has been prepared in two steps as follows: equimolar amounts of ethyl cyanoacetate and ethyl dichlorocyanoacetate were added sequentially to a finely ground suspension of K_2CO_3 in EtOAc at 0°C, then excess acrolein was added and the mixture stirred at room temperature for 3 days. The product from this reaction was treated with PCl_3/HCl in hot DMF, which gave ethyl 2-chloronicotinate in 65% yield.

Elucidate the reaction sequence and give mechanisms for each step.

105. A 2-Arylpropanoic Acid Synthesis

Israeli workers have recently described a very elegant and potentially very useful method for the preparation of 2-arylpropanoic acids, illustrated as follows. Reaction of 3-ethyl-5,6-dihydro-1,4,2-dioxazine with NBS/DBP in the dark gave the expected bromo derivative in *ca.* 80% yield. Addition of this bromo compound to a solution of $AgBF_4$ and 1,4-dimethoxybenzene followed by stirring at room temperature for 18 hours gave, after work up, a product **A**, $C_{13}H_{17}NO_4$, in 90% yield. Heating **A** in $HCl/H_2O/MeOH$ at 65°C for 10 hours gave 2-(2,5-dimethoxyphenyl)-propanoic acid in 90% yield.

Elucidate the overall reaction scheme and give a mechanism for the formation of **A**.

106. Synthesis of Isoquinolin-4-ones

A solution of the nitrile **1** in chloroform was added to 94% sulfuric acid at 0°C and the mixture then stirred at room temperature for 15 minutes. An ice-water quench followed by basic work-up gave a mixture of **2** and **3** in poor yield.

Suggest mechanisms for the formation of **2** and **3**.

2, R^1 = OMe, R^2 = H
3, R^1 = H, R^2 = OMe

107. Naphthopyrandione from 1,4-Naphthoquinone

Reaction of 2-methyl-1,4-naphthoquinone with the ylide derived from phenacylpyridinium bromide in acetonitrile gives 2-methyl-3-phenacyl-1,4-naphthoquinone in 82% yield. Under the same conditions, but using 2-phenoxymethyl-1,4-naphthoquinone instead of the 2-methyl derivative, the product was 3-phenyl-1*H*-naphtho[2,3-*c*]pyran-5,10-dione (97%).

Give mechanistic explanations for these transformations.

108. A Photochromic Product for Sunglasses

Photochromic sunglasses should ideally be colourless under low-light conditions, be converted rapidly to a form with an appropriate absorption spectrum under intense light, and revert quickly to colourless under low light. All of this needs to happen efficiently over a wide temperature range and the system must be capable of an almost infinite number of cycles.

One of the most widely used photochromic compounds for sunglasses is prepared easily as follows: reaction of 2,7-dihydroxynaphthalene with a mixture of sodium nitrite and sulfuric acid at 0°C gives a 98% yield of a dark purple solid. Condensation of this solid with 1,2,3,3-tetramethyl-3H-indolinium iodide in hot ethanol in the presence of triethylamine gives the photochromic product in *ca.* 40% yield, and this can be coated directly on to sunglasses using standard technology.

Deduce the structure of the photochromic compound and explain the basis of its action.

109. A Quick Entry to the Aklavinone Ring System

In a clever, convergent and regiospecific route to the aklavinone ring system, it was shown that treatment of the lithio anion of the cyanophthalide **1** with the enone lactone **2** in THF/HMPA gave the tetracyclic quinone **3** directly (25% yield after conversion to the methyl ester and purification).

Explain this transformation in mechanistic terms.

| 1 | 2 | 3 |

110. A "Stable Enol" that Doesn't Exist

It was reported that the structure of the product obtained (68% yield) by acid-catalysed (8% HCl, reflux, 18 hr) hydrolysis and decarboxylation of the β-keto ester **1** was not the expected diketone, but the stable enol **2** (mp 128-130°C), and that the equilibrium could not be reversed from enol to ketone, even on treatment of **2** with dilute alkali for 10 days. The evidence cited in support of **2** was entirely spectroscopic, as follows: "$C_{14}H_{16}O_3$ (232) : M^+ 232, λ_{max} (ethanol) 266 nm (log ε, 4.09). The IR spectrum of 7 (≡ **2**) showed a broad band between 3000-3500 cm^{-1} (enolic OH) and a sharp peak at 1705 cm^{-1} (ring -C=O); in the NMR spectrum (CDCl₃) of 7 (≡ **2**) a broad peak was present at δ5.95 (1H, olefinic) and a downfield signal (exchangeable with D₂O) at δ11.1 (1H) indicating the presence of an enolic OH group."

Assess this evidence in terms of structure **2**, show that the claim to have prepared **2** as a stable enol is specious, and, on the basis of mechanism, devise a structure for the product from the acid-catalysed hydroysis of **1** (the molecular formula $C_{14}H_{16}O_3$ for the product is correct).

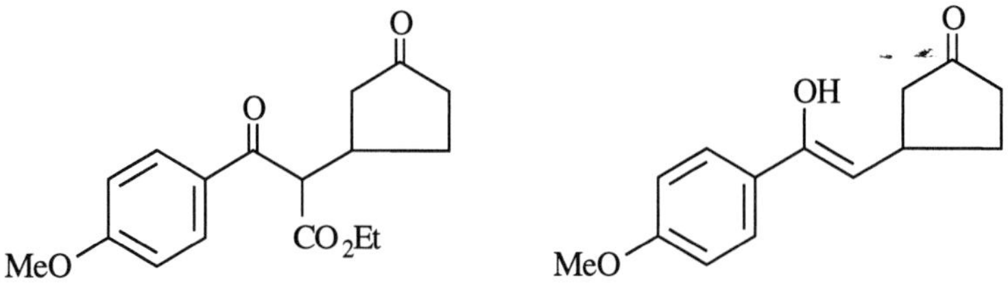

1   **2**

111. Conversion of an Indole-Based Bicyclo[5.3.1]undecane into a Bicyclo[5.4.0]undecane

Treatment of the product derived from reaction of the *N*-oxide **1** with trifluoroacetic anhydride with sodium borohydride gives the tetracyclic indole derivative **2**.

Give a mechanistic explanation for this observation.

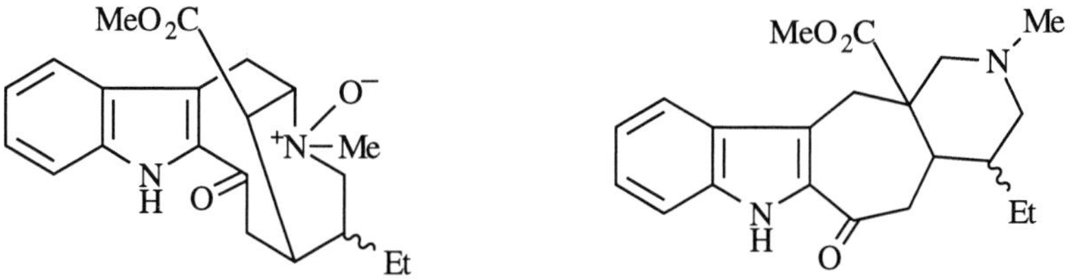

1

2

112. A Classical Mannich Approach to Isoquinoline Alkaloids Leads to an Unexpected Product

Heating a mixture of 1-(2,5-dibromo-3-thienylmethyl)-6,7-dimethoxy-1,2,3,4-tetrahydro-isoquinoline under reflux with aqueous formaldehyde and formic acid for 1 hour gives the tetracyclic compound **1** in 91% yield.

Suggest a mechanism for the formation of **1**.

1

113. Reaction of Benzothiazole with DMAD

Reaction of equimolar amounts of benzothiazole and dimethyl acetylenedicarboxylate in hot, aqueous methanol gives a colourless crystalline solid **A**, $C_{13}H_{13}NO_5S$, in almost quantitative yield. The ir spectrum of **A** (Nujol) shows strong bands at 1745, 1740 and 1690 cm^{-1}, while the ^1H nmr spectrum consists of signals at δ8.9 (1H, s); 7.70 - 7.10 (4H, m); 6.15 (1H, d, $J = 5$ Hz); 4.70 (1H, d, $J = 5$ Hz); and 3.65 (6H, s).

Deduce the structure of **A** and give a mechanism for its formation.

114. Inhibition of the L-DOPA to L-Dopaquinone Oxidation

Oxidation of L-DOPA to L-dopaquinone is an important biological process, and dopaquinone is known to play a key role in the oxidative conversion of tyrosine into melanins, the primary pigments of skin and hair. It has been found that under biomimetic conditions both resorcinol and phloroglucinol inhibit this action of L-DOPA, and the compound **1** was isolated from the enzyme-catalysed L-DOPA/phloroglucinol reaction. Compound **1** could also be prepared by ferricyanide oxidation of a mixture of L-DOPA and phloroglucinol.

Suggest a mechanism for the formation of **1**.

1

115. A One-Pot Synthesis of 1,3,6,8-Tetramethyl-2,7-naphthyridine

There are relatively few synthetic routes to 2,7-naphthyridines; almost all are multistage and rely on annulation of a second pyridine ring to a suitably functionalised pyridine precursor. French workers have described an efficient, one-pot synthesis of 1,3,6,8-tetramethyl-2,7-naphthyridine **1** which involves treatment of a mixture of acetyl chloride (1.6 mol) and aluminium chloride (0.3 mol) with t-butanol or t-butyl chloride (0.1 mol) at 35°C for half an hour, followed by careful addition of the reaction mixture to liquid ammonia. This gave 37% of 2,4,6-trimethylpyridine and 63% of the naphthyridine **1** in a total yield of 91%.

Suggest a mechanism for the formation of **1**.

116. Unexpected Course of a Mannich Reaction in Alkaloid Synthesis

In a projected synthesis of the protoberberine alkaloid thalictrifoline **1** the tetrahydroisoquinoline derivative **2** was treated with formaldehyde and acid under a variety of conditions. Unexpectedly, the product was the isoquinobenzoxazepine derivative **3**, which was obtained in yields of 70-85% depending on the conditions used.

Suggest a mechanism for this transformation.

117. Synthesis of Thioindigo

Thioindigo **1** is obtained in 80% yield when 2,3-dibromothiochromone *S*-oxide is heated with 2 equivalents of sodium acetate in refluxing acetic acid for 15 minutes. No thioindigo is formed if 2,3-dibromothiochromone is used as starting material.

Give a mechanism to account for the formation of thioindigo from the *S*-oxide.

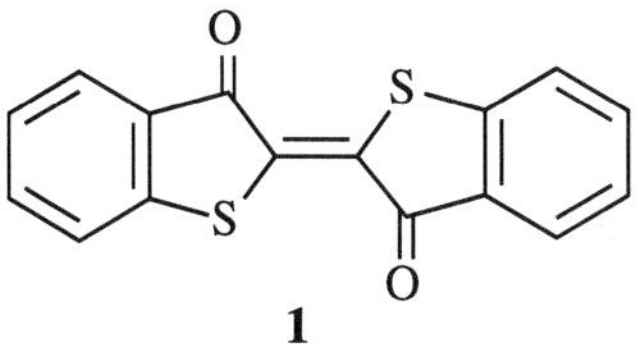

1

118. An Unexpected Reaction During Studies of Amine Oxide Rearrangements

The tetrahydroquinoline derivative **1** was treated with MCPBA in methylene chloride at room temperature in the expectation that the product would be the tricyclic compound **2**. Despite a number of attempts, however, **2** was not formed. The product, obtained in 30% yield, was shown to be **3**.

Explain on a mechanistic basis why **2** was the expected product from the oxidation of **1**, and give a mechanism for the formation of **3**.

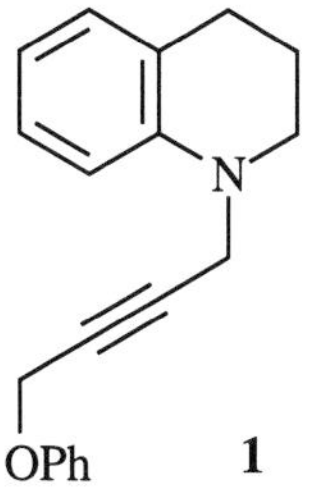

$3\text{-ClC}_6\text{H}_4\text{CO}_2$

1 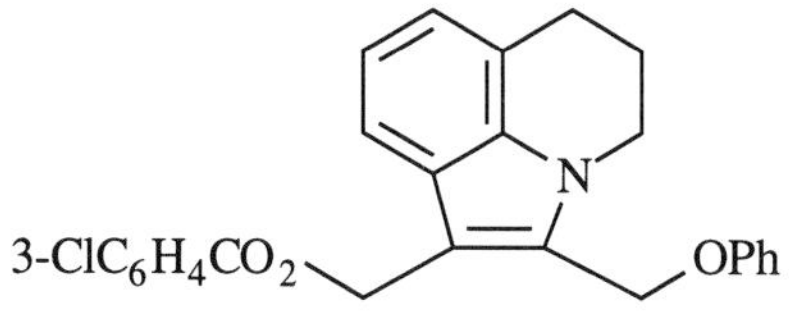 **2** 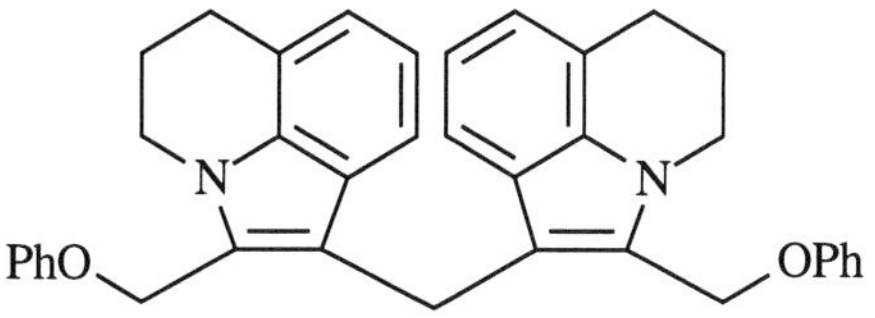 **3**

119. An Unusual Phosphite-Induced Deoxygenation of a Nitronaphthalene

The construction of benzo-fused heterocycles by triethyl phosphite-induced deoxygenation of aromatic nitro compounds is a now classical type of reaction.   Heating of 2-(2-nitrophenoxy)naphthalene with triethyl phosphite under reflux under nitrogen, for example, gave the benzophenoxazone **1** in low yield (unspecified). More interestingly, reaction of the isomeric 1-(2-nitrophenoxy)naphthalene under identical conditions also gave **1**, although in only 5% yield, together with various other products.

Suggest mechanisms to account for the formation of **1** from each of the isomeric ethers.

1

120. A Benzimidazole to Pyrroloquinoxaline Transformation

Treatment of 1-methyl-3-(ethoxycarbonylmethyl)benzimidazolium bromide **1** with DMAD in the presence of triethylamine gives a crystalline product, for which the structures **2** and **3** were suggested by Japanese and Rumanian chemists respectively. Structures **2** and **3** are, however, highly improbable, and the correct structure of the product was shown by an English worker to be **4**.

Suggest a mechanism for the formation of **4**.

121. Acid-Catalysed Benzophenone Isomerisation

Treatment of 6-carboxy-5'-chloro-2'-hydroxy-2-methylbenzophenone with concentrated sulfuric acid gives, after quenching of the reaction mixture with water, an almost quantitative yield of 2-carboxy-5'-chloro-2'-hydroxy-3-methylbenzophenone.

Suggest a mechanism to account for this isomerisation.

122. 2-Oxabicyclo[1.1.0]butanone is not Easily Accessible

Attempts were made to epoxidise the C=C bond of diphenylcyclopropenone and thus obtain the 2-oxabicyclo[1.1.0]butanone derivative. Use of either peracetic acid or hypochlorite was not successful and the starting material was recovered. Oxidation with 25% hydrogen peroxide in a mixture of dioxane and 3M aqueous NaOH at room temperature, however, gave a 77% yield of desoxybenzoin.

Suggest a mechanism for this transformation.

123.   A Simple Synthesis of the Indole Alkaloid Yuehchukene

The alkaloid yuehchukene **1** is a potent anti-fertility agent and has recently been synthesised in 10% yield by treatment of 3-isoprenylindole with silica gel and benzene containing a catalytic amount of trifluoroacetic acid.

Give a mechanistic explanation for this synthesis.

1

124. Isoquinoline Rearrangement

Treatment of 1-(3',4'-methylenedioxybenzyl)-2-methyl-6,7-dimethoxyisoquinolinium iodide with lithium aluminium hydride in ether followed by immediate treatment with 2N hydrochloric acid resulted in formation of an intense red-violet colour which very quickly faded to yellow. Basification of the reaction mixture and extraction with chloroform gave an oil which was reduced by treatment with sodium borohydride in ethanol. This gave a product which was isomeric with, but different from, 1-(3',4'-methylenedioxybenzyl)-2-methyl-6,7-dimethoxy-1,2,3,4-tetrahydroisoquinoline, the expected product from this sequence of reactions.

Suggest a structure for the unexpected product and give a reasonable mechanism for its formation.

125. Multiple Sigmatropic Rearrangements

When 3-(4-*p*-tolyloxybut-2-ynyloxy)[1]benzopyran-2-one **1** was heated under reflux in carefully purified chlorobenzene, the exclusive product, formed in 92% yield, was 1-(*p*-tolyloxymethyl)pyrano[2,3-*c*][1]benzopyran-5(3*H*)-one **2**.   The compound **2** was also the exclusive product when either xylene or ethylbenzene was used as solvent. However, when **1** was heated under reflux in chlorobenzene containing a catalytic amount of AIBN, **2** was obtained in 20% yield together with 80% of 2-methyl-1-(*p*-tolyloxymethyl)furo[2,3-*c*][1]benzopyran-4-one **3**. The compound **3** was the exclusive product when PEG-600, *N,N*-dimethylaniline, pyridine or ethylbenzene containing a catalytic amount of *p*-TsOH was used as solvent, the yields being 90, 90, 80 and 90% respectively.

Give mechanistic interpretations for the above observations.

1   2   3

Ar = 4-MeC$_6$H$_4$

126. A Vilsmeier-Induced Annulation to Benzene

Treatment of 5-allyl-1,2,3-trimethoxybenzene with a mixture of *N*-methylformanilide and phosphorus oxychloride gave the dihydronaphthalene derivative **1** in 58% yield.

Suggest a mechanism for this transformation which is consistent with the following facts:

(a)   When PhN(CDO)Me was used in the reaction, compound **2** was isolated.

(b)   When D_2O was used to decompose the reaction mixture no deuterium was incorporated into the product.

(c)   When PhN(CHO)CD$_3$ was used in the reaction, compound **3** was isolated.

1   **2**

3

127. Isopropylidene to 2-Alkyne Conversion

A very unusual - indeed remarkable - type of reaction was discovered in 1986 [S.L. Abidi, *Tetrahedron Lett.*, **27**, 267 (1986)], namely that treatment of various terpenoids containing a terminal isopropylidene unit with a large excess of sodium nitrite in aqueous acetic acid at 60°C results in *formal* loss of methane and production of a 2-alkyne in moderate to good yield. For example, geraniol, **1**, was reported to give the alkyne **2** in 98% yield. Subsequent investigation of the basic process by American (1987), Japanese (1992) and French (1995) workers established that the overall transformation did indeed occur as originally reported, although yields of alkynes obtained by these other workers were significantly lower than those claimed by Abidi. Clearly, a complex series of reactions in involved.

Suggest a mechanism, or preferably mechanisms, for this transformation.

1   **2**

128. A Tryptamine to Pentacyclic Indoline Transformation

Japanese workers reported in 1956 (*J. Pharm. Soc. Jpn.*, **76**, 966 (1956)) that phosphorus trichloride-catalysed cyclisation of *N*-(3,4-dimethoxyphenylacetyl)tryptamine in boiling benzene gave the expected dihydro-β-carboline derivative. A recent reinvestigation of this reaction, however, showed that the major product, formed in 46% yield, is in fact the spirocyclic indoline derivative **1**. When TFAA in benzene was used for the cyclisation, the indoline **2** was obtained in quantitative yield.

Suggest mechanisms to account for the formation of **1** and **2**.

1   **2**

129. Ring Contractions of a Dibenzothiepinone

Treatment of a cooled (ice-salt bath) solution of the dibenzothiepinone **1** in NMP with sodium hydride gave a deep red solution which was largely decolourised on addition of methyl iodide. Analysis of the reaction mixture showed the presence of unreacted starting material and three new products; the latter were identified as the 6-methyl derivative and the ethers **2** and **3**.

Give mechanisms for the formation of **2** and **3**.

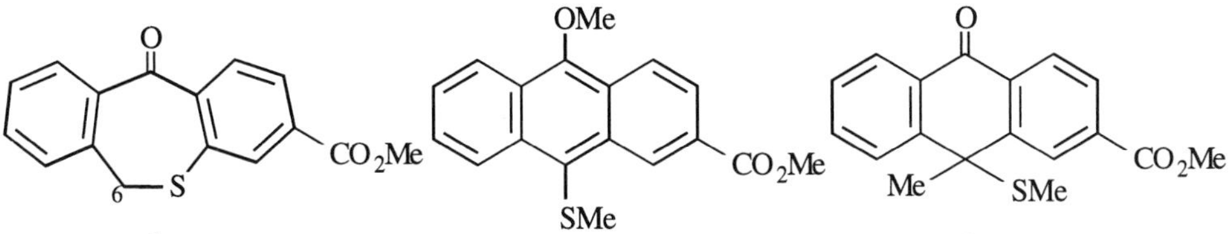

1   2   3

130.   Steroid Fragmentation with DDQ

3β-Acetoxylanost-8-ene **1** was heated with DDQ under reflux in benzene for 24 hours and the crude product subjected to basic hydrolysis followed by chromatographic separation. This gave the three products **2, 3** and **4** in 51,19 and 6% yield respectively. When the reaction time was increased, the amount of **2** obtained decreased, but the amounts of **3** and **4** formed increased.

Suggest mechanisms for the formation of compounds **2, 3** and **4**.

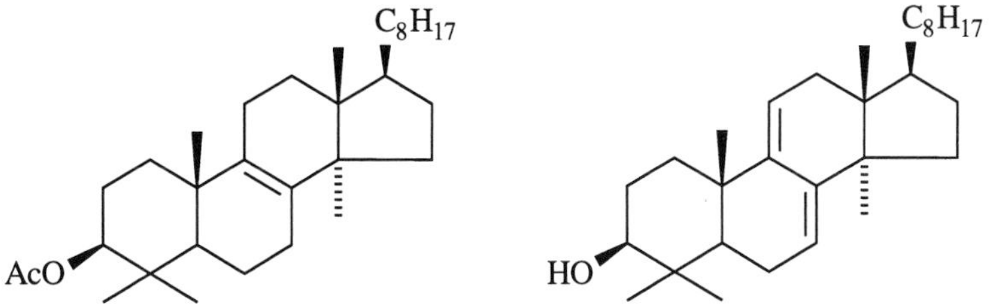

1   2

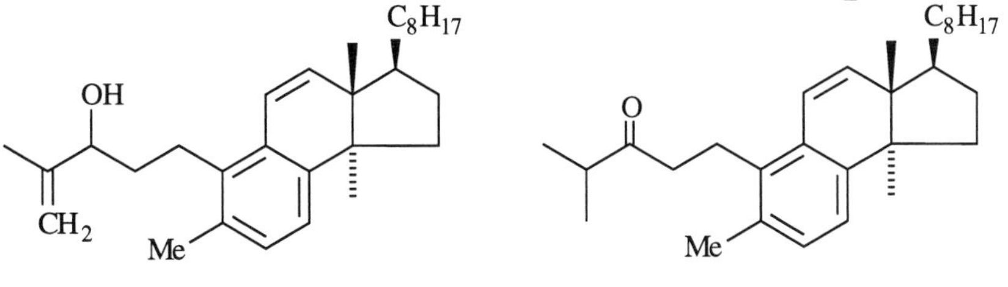

3   4

131. A Base-Induced Vinylcyclopropane Rearrangement

Heating of compound **1** with dimsylsodium in DMSO at 90°C for 21 hours gives a 71% yield of the epimeric ene-diesters **2**.

Suggest a mechanism for this transformation.

132. 4-Demethoxydaunomycin : The Desired Result, but not According to Plan

Subjection of **1** to the sequence of reactions (a) *p*-MeC$_6$H$_4$SH, (b) MCPBA, and (c) aqueous TFA was a key operation in a synthesis of 4-demethoxydaunomycin, and gave **2** in an acceptable 45% yield. The reaction sequence was planned on the basis that (a) **1** would undergo allylic substitution with the thiol; (b) the resulting sulfide would be oxidised by MCPBA to the sulfoxide, which would undergo a [2,3] sigmatropic rearrangement; and (c) hydrolysis would give **2**. Careful examination of the process showed, however, that this pathway was *not* operative.

Devise an alternative mechanistic explanation for the conversion of **1** into **2**.

133. Three Isomeric Hexadienols Give the Same Bicyclic Lactone

Treatment of 1-phenyl-2,4-hexadien-1-ol with the acid chloride of fumaric acid monomethyl ester in ether/triethylamine gave a quantitative yield of the crude ester as an oil. Attempted purification by Kugelrohr distillation at 180-190°C under high vacuum gave the bicyclic lactone **1** in 45% yield. It was subsequently found that the fumarate esters of the isomeric alcohols 1-phenyl-1,4-hexadien-3-ol and 1-phenyl-1,3-hexadien-5-ol also gave compound **1** in 43 and 35% yield respectively when distilled in a Kugelrohr apparatus.

Give reasonable explanations for the formation of **1** from the fumarate esters of the three isomeric alcohols.

1

134. Degradation of Antibiotic X-537A

As part of a chemical study of the antibiotic X-537A, **1**, the latter was treated with 5 equivalents of concentrated nitric acid in glacial acetic acid. The major product from this reaction was a dinitro derivative of **1** which, on treatment with dilute aqueous sodium hydroxide solution, gave a mixture of products one of which was shown to be 6-hydroxy-2,7-dimethyl-5-nitroquinoline.

Give a mechanism for the formation of the quinoline derivative.

1

135.   Steroid Rearrangements : Appearances Can Be Deceptive

Superficially, the following four transformations **1-4** appear to be mechanistically similar.

Show that this is not so, and give a mechanism for each transformation.

136. Synthesis of *Eburna* and *Aspidosperma* Alkaloids

Treatment of the hexahydroindolo[2,3-*a*]quinolizine derivative **1** with dimsyllithium in THF/DMSO gave an excellent yield of the expected diastereomeric β-ketosulphoxides. When the latter were heated under reflux in THF for five minutes in the presence of 6 equivalents of *p*-TsOH, the eburna derivatives **2a** (17%) and **2b** (8%) were obtained together with the two isomeric aspidosperma derivatives **3** (14%) and **4** (17%).

Suggest mechanisms for the formation of **2a, 2b, 3,** and **4**.

1

2a, b

3

4

137. Synthesis of Highly Hindered Cyclic Amines

A recommended procedure for the preparation of highly hindered cyclic amines is illustrated by the following specific example. Reaction of N^1-isopropyl-2,2-dimethyl-1,2-ethanediamine (1 eq.), acetone (2 eq.) and chloroform (1.2 eq.) under standard PTC conditions (CH_2Cl_2, 50% NaOH, 2-5% BTAC, 5°C) gives a quantitative yield of a mixture of 1-isopropyl-3,3,5,5-tetramethyl-2-piperazinone and 4-isopropyl-3,3,6,6-tetramethyl-2-piperazinone in the ratio 73:27. More detailed examination of the reaction showed that careful control of the amount of chloroform used was essential, otherwise secondary reactions occurred. The type of secondary reaction which can occur is illustrated by the quantitative transformation of 2,2,6,6-tetramethyl-4-

piperidone into a mixture of *N*-isopropenyl-5,5-dimethyl-3,3-methylene-2-pyrrolidinone and 3,3-methylene-5,5-dimethyl-2-pyrrolidinone in the ratio 90:10 when treated with excess of chloroform and 50% NaOH in the presence of 18-crown-6. These two products are not interconvertible under the reaction conditions.

Give mechanisms to account for all of the above reactions.

138. A 1,2-Dihydrobenzocyclobutane to Isochroman-3-one Conversion

When 5-methoxybenzocyclobutene-1-carboxylic acid **1** is heated at 150-160°C for 45 minutes the product (48%) is 6-methoxyisochroman-3-one **2**.

Suggest a mechanism for this transformation.

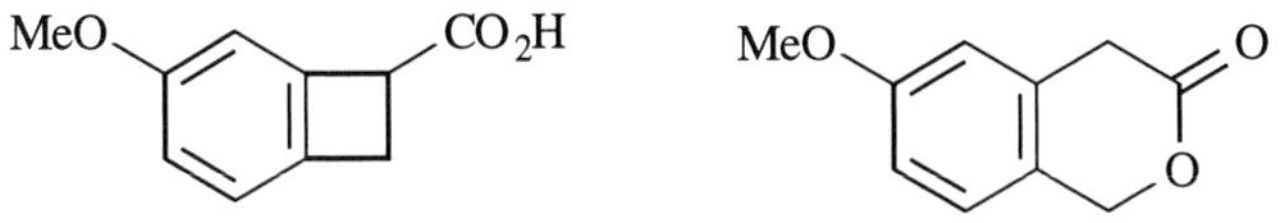

| 1 | 2 |

139. An Unexpectedly Facile Decarboxylation

Hydrolysis of 1,5-diethoxycarbonyl-3-methyl-3-azabicyclo[3.3.1]non-9-one with hot 20% aqueous hydrochloric acid proceeds smoothly and gives 3-methyl-3-azabicyclo[3.3.1]non-9-one in 70% yield.

Give a mechanism for this transformation which does not involve anti-Bredt intermediates.

140. Regiospecific Synthesis of a 1,2-Dihydronaphthalene

Brief treatment of the sulfoxide **1** with trifluoroacetic anhydride in refluxing toluene gives the dihydronaphthalene **2** in 55% yield as a single regioisomer

Give a mechanism for this transformation.

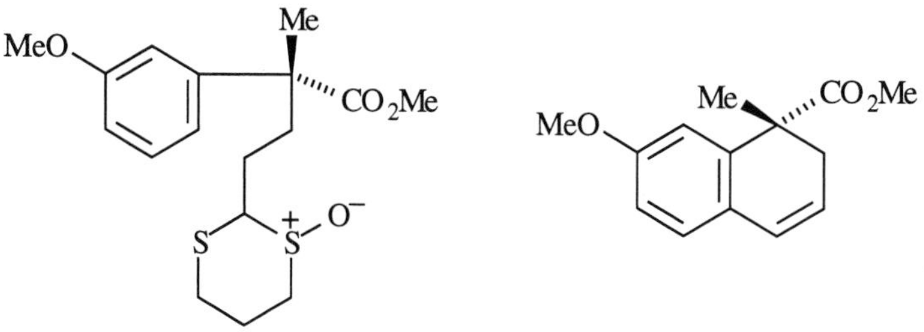

1   **2**

141. Degradation of Terramycin

During investigation of the structure of the antibiotic terramycin **1** a curious reaction was observed. Thus, when terramycin was treated with dilute base, the substituted indanone **2** was formed among many other products.

Devise a mechanism for the formation of **2** from terramycin.

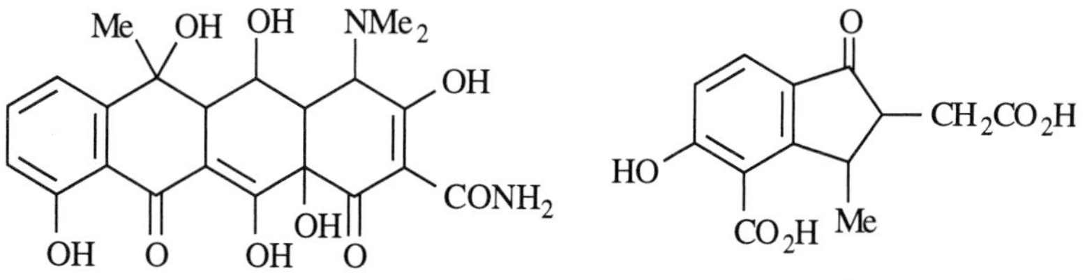

1

Terramycin

2

142. Synthesis of Bi-indane-1,3-dione

Reaction of 2,3-dichloro-1,4-naphthoquinone with one equivalent of the anion of indane-1,3-dione at room temperature gave an excellent yield of the expected substitution product. Brief exposure of this product to ethanolic potassium hydroxide solution resulted in smooth, high yield substitution of the remaining chlorine atom. Monobromination of this latter product with bromine in acetic acid proceeded in quantitative yield, and treatment of the bromide with ethanolic potassium hydroxide solution gave "bi-indane-1,3-dione" **1**.

Deduce the reaction sequence and give mechanisms for the various transformations.

1

143. A Coumarin to Cyclopentenone Transformation

Treatment of 3,4-phenacylidene-3-acetylcoumarin, which is easily prepared by reaction of 3-acetylcoumarin with phenacyl bromide in the presence of sodium ethoxide, with aqueous sodium hydroxide results in smooth conversion into 4-(2-hydroxybenzylidene)-3-phenylcyclopent-2-en-1-one.

Give an explanation for this transformation.

144. "Synthetic Heroin" and Parkinsonism

The prototype of the 4-phenylpiperidine analgesics is **1** ("pethidine", "meperidine", "demerol"), which is widely used in clinical medicine, especially for relief of pain during childbirth (unlike more potent analgesics it induces little depression of respiration in the baby). It is, however, potentially addictive and is widely abused. Standard SAR studies have shown that so-called "reversed meperidine", **2**, which is obviously much easier to synthesise than **1**, has similar analgesic and addictive properties to **1**, and compound **2**, often referred to in drug sub-culture as "synthetic heroin", is also widely abused. However, some care must be taken during the preparation of **2** in the acetylation of the tertiary alcohol derived from reaction of 1-methyl-4-piperidone with phenylmagnesium bromide, otherwise elimination of water and/or acetic acid can occur and lead to significant amounts of the olefin **3**.

It has been shown that **3** is biotransformed in a reaction catalysed by monoamine oxidase B to species that cause the selective degeneration of nigrostriatal neurons, giving rise to a Parkinsonian syndrome in man and other primates. Studies of this process have shown that the pyridinium salts **4** and **5** are involved, and that **4** undergoes spontaneous disproportionation to **5** and **3**; **5** is the putative ultimate neurotoxin. More recent studies have shown that **4** undergoes a spontaneous reaction in pH 7.4 buffer to give methylamine and a product identified as **6**.

Give a mechanistic interpretation for the formation of **6**.

1, R = CO₂Et
2, R = OCOMe

3

4

5

6

145. Rearrangement of 6-Hydroxyprotopine to Dihydrosanguinarine

Sanguinarine **1** is the prototype of benzophenanthridine alkaloids which occur frequently in plants of the family *Papaveraceae*. It has anticancer activity and has become of commercial interest because of its antibacterial properties. It is known from biosynthetic studies that protopine **2** is a metabolite on the pathway to sanguinarine, and recent work has shown that a microsomal cytochrome P-450 NADPH dependent enzyme will hydroxylate protopine specifically at C-6.   The resulting 6-hydroxyprotopine rearranges spontaneously to dihydrosanguinarine **3**.

Suggest a mechanistic explanation for the formation of **3** from 6-hydroxyprotopine.

1   2

3

146.   Biotin Synthesis : Sulfur Preempts a Beckmann Rearrangement

During an elegant synthesis of biotin the oxime **1** was treated with thionyl chloride at 0°C in the expectation that Beckmann rearrangement would occur. The product, however, was found to be **2** (75%).

Suggest a mechanism for the formation of **2**.

1   2

147. A 1,4-Dithiin to Thiophene Rearrangement

Methylation of 3-nitro-2,5-diphenyl-1,4-dithiin using a mixture of methyl iodide and silver tetrafluoroborate in methylene chloride/acetonitrile/nitromethane gave a mixture of two products in the ratio of 3:1. The major product was shown to be 1-methyl-6-nitro-2,5-diphenyl-1,4-dithiinium tetrafluoroborate and the minor component to be 3-(dimethylsulphonio)-5-nitro-2,4-diphenylthiophene tetrafluoroborate.

Give a mechanistic explanation for formation of the latter compound.

148. An Intramolecular Wittig Reaction

Reaction of 4-(2-formylphenoxybutyl)triphenylphosphonium bromide with sodium ethoxide in refluxing DMF gave a mixture of 3,4-dihydro-2*H*-1-benzoxocin and 2-ethyl-2*H*-1-benzopyran in the ratio 95:5 and in 31% yield.

Suggest mechanisms to account for the formation of these products.

149. Amination of a Cephalosporin

Treatment of the β-lactam **1** with *N*-methylhydroxylamine gave the expected condensation product, pyrolysis of which in toluene at 110°C led to formation of the isomers **2** in 72% yield.

Give an explanation for the formation of **2**.

1   **2**

150. A Uracil to 1,2,3-Triazole Conversion

Hydrolysis of 5-diazo-1-methyluracil-6-methanolate **1** with 5% (v/v) aqueous acetonitrile at 100°C gives 1-methyl-4-carboxamido-1,2,3-triazole in 78% yield.

Give an explanation for this transformation.

1

151. Schumm Devinylation of Vinyl Porphyrins

In 1928, Schumm discovered that heating vinyl porphyrins in molten resorcinol resulted in facile loss of the vinyl groups from the porphyrins. It was subsequently shown that 1'-hydroxyalkyl, formyl and acetyl porphyrins behave similarly under the same conditions, i.e. these substituent groups also are removed. The Schumm devinylation reaction has been used in both synthetic and degradation studies on porphyrins, and it was stated in a paper in the Journal of Organic Chemistry that "One of the most useful reactions in the chemistry of natural porphyrins is the devinylation of vinyl porphyrins in molten resorcinol . . . . "

Suggest a mechanism for the Schumm devinylation reaction.

152. A Uracil to Substituted Benzene Transformation

Refluxing of 1,3-dimethyl-5-formyluracil with acetylacetone in ethanolic sodium ethoxide for 2 hours followed by acidification of the mixture with concentrated hydrochloric acid gives ethyl 3-acetyl-4-hydroxybenzoate in 55% yield, together with 1,3-dimethylurea.

Japanese workers have studied this transformation, and postulated two very reasonable mechanisms. When the deuterated uracil Het-CDO was used, the product was ethyl 3-acetyl-2-deuterio-4-hydroxybenzoate, in full agreement with one of their suggested mechanisms. Their alternative mechanism would have led to the isomeric ethyl 3-acetyl-6-deuterio-4-hydroxybenzoate.

Identify a reaction mechanism which is consistent with the results of the labelling study.

153. γ-Lactones from Vinyl Sulfoxides

Reaction of vinyl sulfoxides with dichloroketene has been found to result in a new general synthesis of γ-lactones. With phenyl (*E*)-β-styryl sulfoxide, for example, the product, obtained in 65% yield, is the lactone **1**.

Suggest a mechanism for this transformation.

1

154. Tetrahydrothieno[2,3-*c*]- and -[3,2-*c*]pyridine Synthesis

Treatment of *N*-(2-hydroxy-2-phenylethyl)-2-aminomethylthiophene with polyphosphoric acid results in formation of 4-phenyl-4,5,6,7-tetrahydrothieno[2,3-*c*]pyridine (**1**) in 89% yield. If trifluoroacetic acid is used, however, the product, obtained in 100% yield, is a 4:1 mixture of 7-phenyl-4,5,6,7-tetrahydrothieno[3,2-*c*]pyridine **2** and **1**.

Give a mechanistic explanation for these observations.

1   **2**

155.   Oxothiolan → Oxathian → Oxothiolan → Oxathian

Treatment of the oxathiolan **1** with two equivalents of sulfuryl chloride in carbon tetrachloride at 0°C gives the 1,4-oxathian **2** which, on treatment with two equivalents of triethylamine followed by hydrolysis with wet silica gel, is converted into the oxathiolan **3** in 95% yield. The latter is unstable to acid, and when heated in benzene with *p*-toluenesulfonic acid is slowly converted into **4** in 83% yield.

Suggest mechanisms for the formation of **2**, **3** and **4**.

| **1** | **2** | **3** | **4** |

156.   A Thiazolopyrimidine to Pyrrolopyrimidine Transformation

Heating the thiazolo[5,4-*d*]pyrimidine obtained by triethylamine catalysed condensation of 6-chloro-1,3-dimethyl-5-nitrouracil with methyl thioglycolate with an equimolar amount of DMAD in 5% aqueous methanol gives the pyrrolo[3,2-*d*]pyrimidine **1** in 71% yield.

Give a mechanism for the formation of **1**.

1

157. Oxidation of 1,5-Diacetoxynaphthalene with NBS

Treatment of a solution of 1,5-diacetoxynaphthalene in acetic acid with four equivalents of NBS in aqueous acetic acid gives 5-acetoxy-2-bromo-1,4-naphthoquinone in better than 90% yield.

Suggest a mechanism for this transformation.

158. Badly Chosen Reaction Conditions?

In a standard hydrolysis procedure the ketol acetate **1** was heated under reflux with 5% methanolic potassium hydroxide solution. The product (60-75% yield) was not the α-hydroxy ketone, however, but was shown to be the tetrahydronaphthol **2**.

Suggest a mechanism for this transformation.

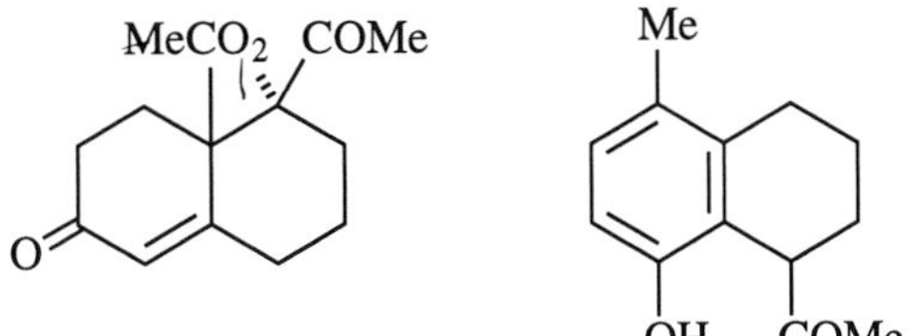

1   **2**

159. One Step Synthesis of a Highly Symmetrical Hexacyclic System from a Simple Naphthol

In an attempt to use phenolic Mannich bases in a synthesis of 3-chromanones, a mixture of 1-[(dimethylamino)methyl]-2-naphthol and α-chloroacrylonitrile was heated under reflux in anhydrous dioxane. The product, however, which was formed in 55% yield, was shown to be the naphthonaphthopyranopyran **1**.

Suggest a mechanism for the formation of **1**.

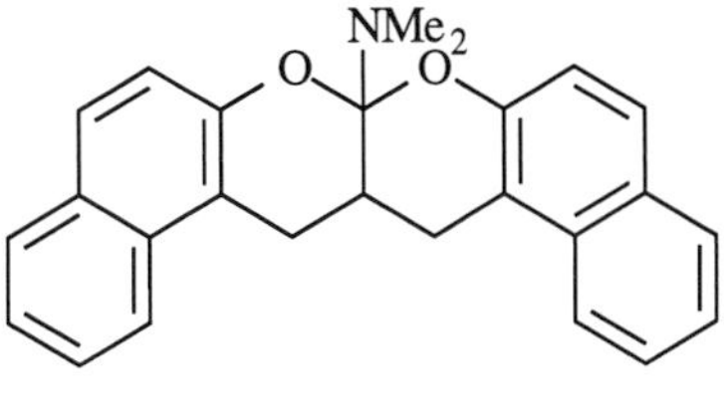

1

160.   A Method for Aryloxylation of *p*-Cresol

Treatment of the sulfoxide prepared from *p*-cresol and thionyl chloride with a mixture of TFAA and KHCO$_3$ in acetonitrile at -30°C led to formation of an orthoquinone mono(monothioketal). Desulfurisation with NiCl$_2$(PPh$_3$)$_2$ gave 2-(4-methylphenoxy)-4-methylphenol in 38% overall yield.

Elucidate the reaction sequence.

161.   Nitriles from *N*-Chlorosulfonylamides

Reaction of π-excessive heterocycles (e.g. thiophene, indole), enol ethers (e.g. dihydropyran) and enol acetates, and carboxylic acids with chlorosulfonyl isocyanate leads in generally excellent yields to *N*-chlorosulfonylamides. These intermediates are converted into the corresponding nitriles by heating in DMF, although the yields can be somewhat variable. A recent reinvestigation of the *N*-chlorosulfonylamide to nitrile conversion revealed that treatment of the amides with one equivalent of triethylamine led to formation of the nitriles in excellent yield. Clearly, the mechanisms of the DMF and the Et$_3$N induced transformations are different.

Suggest mechanisms for each of the transformations

162.   Synthesis of Perylenequinone

The perylenequinone natural product calphostin D **3** is of much current interest due to its potent inhibition of protein kinase-C. The key step in an elegant and very short total synthesis of **3** was the treatment of the *o*-quinone **1** with TFA and then TTFA, which gave **2** in 91% yield.

Give a mechanistic interpretation for the conversion of **1** into **2**.

2, R^1 = Me; R^2 = H; R^3 = Ac
3, R^1 = R^3 = H; R^2 = Me

163.   A Route to 2-Vinylindoles

2- and 3-Vinylindoles are valuable precursors to many carbazoles and to indole alkaloids. The 3-substituted compounds are readily accessible, but the 2-vinyl derivatives have only recently become easily available as a result of a clever synthesis developed by Indian workers. This is illustrated by the conversion of *N*-(allenylmethyl)-*N*-methylaniline **1** into 2-ethenyl-*N*-methylindole **2** in 80% yield by treatment with MMPP in aqueous methanol at room temperature.

Give a mechanistic explanation for the transformation of **1** into **2**.

164.   Brain Cancer : The Mechanism of Action of Temozolomide

Temozolomide **1** is claimed to be the most effective drug developed so far for the treatment of brain cancer. Investigation of its mode of action has revealed that **1** is a prodrug, and that the methyl cation is the active agent in the chemotherapy.

Show mechanistically how methyl cations can be generated from temozolomide **1**.

1

165.   Attempted Nef-Type Reaction Leads to 3-Arylpyridine Synthesis

An unexpected, but potentially useful, reaction was discovered during attempts to effect Nef-like transformations of nitrobicyclo[2.2.1]heptenes into bicyclo[2.2.1]heptenones. Thus, treatment of 6-*exo*-aryl-5-*endo*-nitrobicyclo[2.2.1]hept-2-enes with $SnCl_2.2H_2O$ in hot THF or dioxane led directly to 3-arylpyridines in moderate yield.

Suggest an explanation to account for this transformation.

166.   The ArCOMe → ArC≡CH Transformation

Hydration of terminal alkynes to methyl ketones is a valuable synthetic transformation. The reverse conversion, *viz.* from a methyl ketone to an alkyne can also be a useful transformation. One of the most effective (but largely neglected) procedures for the conversion ArCOMe→ArC≡CH has been shown to be the reaction of the acetophenone with two equivalents of $POCl_3$ in DMF followed by treatment of the intermediate thus produced with a hot solution of sodium hydroxide in aqueous dioxane. The method can be used to prepare mono-, di- and triethynylarenes in 30-50% yield.

Explain this ArCOMe→ArC≡CH transformation in mechanistic terms.

167.   β-Amino Nitriles from Azetidones

Treatment of 4-alkoxyazetidin-2-ones **1** with a catalytic amount of TMSOTf in acetonitrile at 0°C results in smooth conversion to β-amido nitriles **2** in high yield (the *anti* isomer is always the major product).

Suggest a mechanism for this transformation.

1	2

$$R = H, alkyl$$
$$R^1 = Me, allyl$$

168.   A New Route to 1,3-Disubstituted Naphthalenes

Synthesis of 1,3-disubstituted naphthalenes is a non-trivial exercise. In an interesting approach to such compounds it was found that they could be assembled by a simple two-step operation as follows. Condensation of 2-(trifluoromethyl)benzaldehyde with pentan-3-one gave the expected aldol product as the (*E*)-isomer, treatment of which with lithium 4-methylpiperazide in ether at -10°C led directly to 1-methylpiperazino-3-propionylnaphthalene in 30% yield.

Suggest a mechanism for the annulation reaction.

169. An Efficient Annulation Route to 6-Substituted Indoles

During synthetic studies on the antitumour antibiotic CC-1065, ready access was required to a series of 6-substituted indole-3-carboxylates. The following example illustrates the successful general strategy that was developed: reaction of the N-Cbz derivative of N-(3-methylphenyl)hydroxylamine with methyl propiolate in nitromethane as solvent and in the presence of Hünig's base gave methyl 1-benzyloxycarbonyl-6-methylindole-3-carboxylate directly in 89% yield.

Suggest a mechanism to account for this transformation.

170. Failure of a Rearrangement : from a Useful Compound to a Useless Product

During studies on pyrethroid insecticides methyl permethrate **1** (a mixture of *cis* and *trans* isomers) was pyrolysed at 260-270°C in the expectation that a vinylcyclopropane-to-cyclopentene rearrangement would occur. The product, however, was found to be methyl *o*-toluate (78%).

Suggest a mechanistic explanation for this unexpected transformation.

1

171. Synthesis of Trisubstituted Isoxazoles

1-Ethoxyethene undergoes smooth, high yielding regioselective condensation with 2-benzoylbut-1-en-3-one, and treatment of the product with 2 or 3 equivalents of hydroxylamine hydrochloride in refluxing ethanol gives the isoxazole **1** in 99% yield.

Explain the reaction sequence and suggest a mechanism for the formation of **1**.

1

172.   2,3-Dihydrobenzofurans from 1,4-Benzoquinones

Researchers in Chile have recently shown that treatment of the product obtained from condensation of 2-cyano-1,4-benzoquinone with *(E)*-1-trimethylsilyloxybuta-1,3-diene with dilute hydrochloric acid results in smooth, high yielding conversion to the dihydrobenzofuran **1**.

Suggest a mechanism for the acid catalysed rearrangement.

1

In closely related studies, French workers reacted 1,4-benzoquinone with the *N,N*-dimethylhydrazone of propenal, using BF$_3$.OEt$_2$ as catalyst. Treatment of the initially formed product with copper(II) acetate in buffered aqueous THF gave 2-formyl-5-hydroxy-2,3-dihydrobenzofuran in excellent overall yield.

Explain this sequence of reactions in mechanistic terms.

173.   Easy Construction of a Tricyclic Indole Related to the Mitomycins

As part of a programme of studies on the synthesis of mitomycins, the dione **1** was treated with sodium hydride in dry THF at room temperature. This gave an epimeric mixture of carbinolamines which was highly acid sensitive and which, on treatment with glacial acetic acid at room temperature for 5 minutes, underwent dehydration to the tricyclic indole derivative **2**.

Suggest a mechanism for the transformation of **1** into **2**.

1   **2**

174. Cycloaddition to a Benzothiopyrylium Salt

Treatment of the 2*H*-1-benzothiopyran **1** with trityl fluoroborate in nitromethane at room temperature gave the corresponding benzothiopyrylium salt in 90% yield as an unstable green solid which reacted smoothly with 2,3-dimethylbuta-1,3-diene at room temperature to give the expected cycloadduct in 94% yield. Addition of the cycloadduct to an ice-cold solution of triethylamine in ethanol resulted in formation of **2** and **3** in 51 and 28% yield respectively, the latter as a 1:1.2 mixture of diastereomers. Mixtures of **2** and **3** were also obtained if other bases/solvent systems were used, and careful nmr study of the base-induced transformations clearly revealed that an unstable intermediate was being formed.

Elucidate the overall reaction scheme, suggest a structure for the unstable intermediate, and give mechanisms for the formation of **2** and **3** which are consistent with the further observation that when oxygen was bubbled through the triethylamine/cycloadduct mixture, the yield of **2** decreased from 51 to 7% while that of **3** increased from 28 to 73%.

1   **2**   **3**

175. Benzotriazole from 1,2,4-Benzotriazine *N*-Oxides

Both the 1- and 2-oxides of 3-methyl-1,2,4-benzotriazine can be obtained by direct oxidation of the parent heterocycle with peracid, and both give benzotriazole in low yield on brief treatment with 2% aqueous NaOH.

Suggest mechanisms for the ring contraction reactions.

176. A Furan → Furan Transformation

Treatment of **1** with morpholine at room temperature results in an exothermic reaction and the reaction mixture turns dark red in colour. Addition of water results in the separation of **2** as yellowish-green needles in 54% yield.

Suggest a mechanism for the transformation of **1** into **2**.

| 1 | 2 |

177. Ring Expansion of Both Rings of Penicillin Sulfoxides

A number of penicillin sulfoxides **1** were heated in THF with two equivalents of ethoxycarbonyl isocyanate in the expectation that the corresponding sulfilimines would be obtained. The products, however, were shown to be the imidazo[5,1-*c*][1,4]thiazines **2**, which were obtained in 31-54% yield.

Give a mechanistic interpretation for this transformation.

| 1 | 2 |

178. A Retro-Pictet-Spengler Reaction

The Pictet-Spengler route to tetrahydro-β-carbolines is frequently used in indole alkaloid synthesis, and much attention has been devoted in recent years to the development of enantiospecific Pictet-Spengler reactions and the factors which influence the diastereochemistry at C-1 and C-3 (*c.f.* **1**). The diester **1a** was prepared from **2** as 1:1 mixture of diastereomers and heated in 2% ethanolic hydrogen chloride for 3 hours in an attempt to effect epimerisation at C-1 and increase the amount of *trans* isomer. The product, however, was **2**, which was isolated in 76% yield. Heating of the *cis* and *trans* diesters **1** separately in ethanolic hydrogen chloride for 3 hours also gave **2**. By comparison, when a *cis*, *trans* mixture (39:61) of **1b** was stirred in a mixture of methylene chloride and trifluoroacetic acid at room temperature for 90 minutes the *trans* diester was obtained in 96% yield.

Suggest mechanisms to account for the degradation of **1** into **2**, and for the epimerisation of *cis*-**1b** to *trans*-**1b**.

1a, R = SO$_2$Ph
b, R = Me

2

179. An Olefin to α-Hydroxy Ketone Transformation

Oxidation of the olefin **1** with an excess of monoperphthalic acid in ether/dichloromethane at 4°C in the dark gave a mixture of epimeric ketols **2** in 74% yield (*trans*:*cis* 6:1).

Suggest a mechanism for this transformation.

1

2

180. A Benzofuran from a Cyclopropachromone

Treatment of a solution of the cyclopropachromone **1** in DMSO with a solution of dimethylsulfonium methylide in DMSO/THF at room temperature for 1.5 hours followed by quenching of the reaction mixture with water gave the phenacylbenzofuran **2** in about 30% yield.

Devise a mechanism for this transformation.

181. A Pyridocarbazole to Pyridazinocarbazole Rearrangement

Attempts to reduce the ketonic function of the tetrahydrocarbazole derivative **1** under standard Wolff-Kishner conditions led, unexpectedly, to formation of the pyridazinocarbazole **2** in 62% yield.

Suggest a mechanism for this transformation.

182. Substituent Group Effect During Sulfide → Sulfoxide Oxidation

Attempts to oxidise the sulfides **1a-e** to the corresponding sulfoxides using 30% hydrogen peroxide in glacial acetic acid were unsuccessful, and resulted only in formation of the corresponding sulfones. When 30% hydrogen peroxide in a mixture of acetic acid (60%) and water (40%) was used, however, the sulfides **1a-d** were smoothly oxidised to the corresponding sulfoxides. Oxidation of **1e** under these conditions gave *p*-nitrobenzaldehyde in 90% yield.

Suggest a mechanism to account for the anomalous behaviour of **1e**.

CH$_2$SPh

X

1 a, X = MeO; **b**, X = H; **c**, X = Cl; **d**, X = CN; **e**, X = NO$_2$

183.   Ring Contraction of a Benzodiazepine

Refluxing of a mixture of 2-amino-7-chloro-5-phenyl-3*H*-1,4-benzodiazepine and 2,4-pentanedione (no solvent) gives, among other products, an 11% yield of 3-acetyl-7-chloro-2-methyl-9-phenyl-1*H*-pyrrolo[3,2-*b*]quinoline.

Devise at least two acceptable mechanisms for this transformation.

184.   An Indane-1,3-dione Synthesis

Condensation of 3-nitrophthalic anhydride with 2,4-pentanedione in the presence of pyridine and piperidine followed by acidification gives 2-acetyl-4-nitroindane-1,3-dione in 76% yield.

Suggest a mechanism for this transformation.

185.   An Azepine to Cyclohexadienone Ring Contraction

2,5,7-Trimethyl-3,4,6-triphenyl-3*H*-azepine was heated under reflux in glacial acetic acid for 4 hours. A complex mixture of products was obtained, from which only one pure product could be obtained in low yield (12%), namely 4,6,6-trimethyl-2,3,5-triphenylcyclohexa-2,4-dienone.

Suggest a mechanism for formation of the cyclohexadienone.

186.   Xanthopterin from Pterin 8-Oxide

A suspension of pterin 8-oxide **1** in a 1:1 mixture of TFA-TFAA was stirred for 1 hour at 50°C. The solvent was then evaporated and the residual solid hydrolysed with ammonium hydroxide/sodium hydroxide solution. Acidification gave xanthopterin **2** in quantitative yield.

Account for the **1** to **2** transformation in mechanistic terms.

187.   Fused Dihydro-1,4-dithiins from Chromanones

An excellent method for the preparation of *gem*-difluoro compounds from aldehydes and ketones consists of conversion of the carbonyl compound to the corresponding 1,3-dithiolane followed by treatment with two equivalents of 1,3-dibromo-5,5-dimethylhydantoin (DBH) and pyridinium poly(hydrogen fluoride) (HF-pyridine) in methylene chloride. Attempted extension of this procedure to 7-methoxy-2,2-dimethyl-4-chromanone, however, gave only the dihydro-1,4-dithiin derivative **1** in 78% yield. This transformation, which proceeded in excellent yield with a variety of 4-chromanones, was found to require only the DBH (i.e. fluoride ion played no role).

Suggest a mechanism for the formation of **1**.

188. The "Additive Pummerer Reaction"

Vinylic sulfoxides such as **1** react readily with electrophiles to give highly reactive species, and the overall reactions have been likened to "generation" of the synthon **2**. Treatment of **1** with TFAA, for example, results in what is referred to as an "additive Pummerer reaction", and gives the diester **3**. Reaction of **1** with triflic anhydride and sodium acetate in acetic anhydride, by contrast, gives an 85% yield of the protected aldehyde **4**.

Based on the concepts of synthon **2** and the "additive Pummerer reaction" suggest a mechanism for the **1** → **4** transformation.

| 1 | 2 | 3 | 4 |

189. An Illustration of the Problem of Artefacts in Natural Product Chemistry

A perennial problem in structural elucidation of natural products is to decide/confirm that any pure compound obtained from isolation and purification techniques is a genuine natural product and not an artefact of the isolation and/or purification methods employed. A recent example illustrates the problem well. The "alkaloid" magallanesine was shown to have the structure **1**, but it is difficult to account for this structure in biosynthetic terms. Shamma reinvestigated the problem and found that simple chromatography of oxyberberine **2** on silica using chloroform as eluent gives magallanesine in excellent yield, thus indicating that **1** is almost certainly an artefact and not a naturally occurring compound.

Give a mechanistic explanation for the conversion of **2** into **1**.

| 1 | 2 |

190.   Vinylogy, and a Stereochemical Puzzle

Treatment of 1,1,1-trichloro-2-penten-4-one with aqueous sodium hydroxide solution gives a good yield of 5-chloro-*trans*-2-*cis*-4-pentadienoic acid after standard acid work up.

Suggest a mechanism for this transformation. (Note : the original investigators were unable to account satisfactorily for the observed stereochemistry.)

191.   A Simple, High Yielding Route to a Cage Compound

Heating a solution of the triketone **1** in DME containing a catalytic amount of *p*-TsOH under reflux for 48 hours results in quantitative isomerisation to **2**. Treatment of a THF solution of **1** with t-BuOK/t-BuOH at room temperature for 16 hours, on the other hand, gives a 78% yield of the diketone **3**.

Suggest a mechanism for the formation of **3**.

HC(CH₂COCH₂Cl)₃		
1	**2**	**3**

192.   Not all Ketals Hydrolyse Easily in the Expected Manner

Various attempts to hydrolyse 9-ethylenedioxybicyclo[3.3.1]nonane-3,7-dione under acid conditions resulted only in the formation of either 5-hydroxyindan-2-one (HCl/AcOH; 14% yield) or a mixture of the indanone (13%) and 7,9-bis(ethylenedioxy)bicyclo[3.3.1]nonan-3-one (13%; 10% aqueous HCl).

Give an explanation for the formation of these products.

193. From a Tricycle to a Ring Expanded Bicycle

During studies directed towards a total synthesis of gibberellic acid, the lactone **1** was heated under reflux with potassium t-butoxide in t-butanol. Acidification of the reaction mixture and separation of the products gave compound **2** as the major product (30%).

Give a reasonable mechanism for the conversion of **1** into **2**.

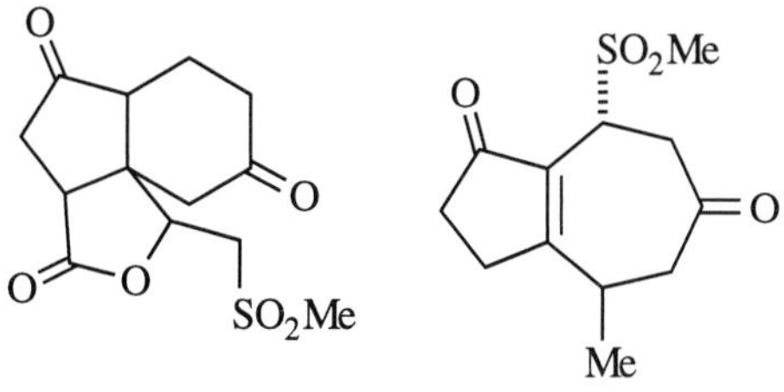

1	**2**

194. A Flavone from a Chromanone

Treatment of the chroman-4-one **1** with benzyl chloride in DMF at 100°C gave the corresponding benzyl ether **2**. When the reaction temperature was raised to 153°C, however, the products obtained were the chromone **3** and the flavone **4**. It was subsequently shown that the same type of rearrangement could be effected simply by heating **1** with benzyl chloride in DMF containing potassium carbonate.

Suggest a mechanism for the formation of **4**.

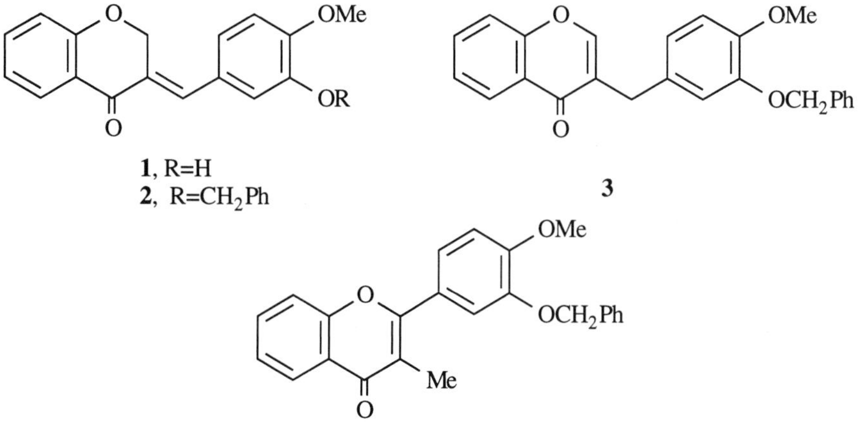

1, R=H
2, R=CH₂Ph

3

4

195. Thiophene Ylide Rearrangement

Thermolysis of the ylide **1** in anisole containing a catalytic amount of $BF_3.OEt_2$ gives a mixture of products, the major component of which (35%) is **2**.

Give a mechanism for the formation of **2**.

1   2

196. A Stepwise $2\pi + 2\pi$ Intermolecular Cycloaddition

Reaction of 1,2-dimethylcyclohexene with the ethylene glycol acetal of acrolein in methylene chloride in the presence of 25 mol % of $BF_3.OEt_2$ at -78 to -10°C for 2 hours gives a 70% yield of the cycloadduct **1** in a formal $2\pi + 2\pi$ intermolecular cycloaddition. All of the evidence for this and related reactions, however, indicates a stepwise mechanism for the formation of **1**.

Suggest the stepwise mechanism.

1

197. Rearrangement of an Aryl Propargyl Ether

The propargyl ether **1** was heated in refluxing dimethylaniline for 5 hours in the expectation that Claisen rearrangement followed by cyclisation would lead to the linear tricyclic system. The only product - isolated in only 12.5% yield - was, however, the angular tricyclic ketone **2**.

Suggest a mechanism for the formation of **2**.

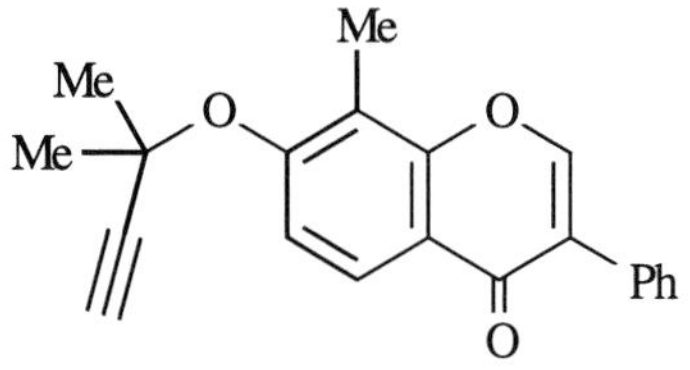

1 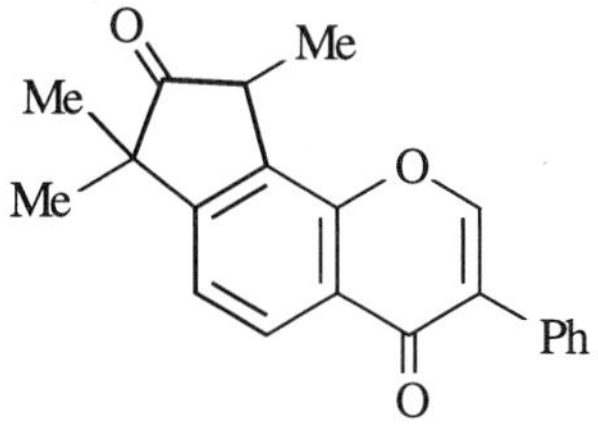 **2**

198. The Best Laid Plans. . . . . .

Many of the most elegant syntheses of complex natural products are based on mechanistic speculation, as illustrated by the following postulated entry to strychnine-based alkaloids. Thus, it was reasoned that treatment of the indole **1** with methyl chloroformate and a non-nucleophilic base followed by acid catalysed rearrangement would lead to the tetracycle **2**. In practice, the first step worked well when Hünig's base was used, but the second step, the acid catalysed rearrangement, failed to give **2**. The only product, isolated in good yield, was the carbazole **3**.

Elucidate the proposed reaction sequence from **1** to **2** and suggest an explanation for the formation of **3**.

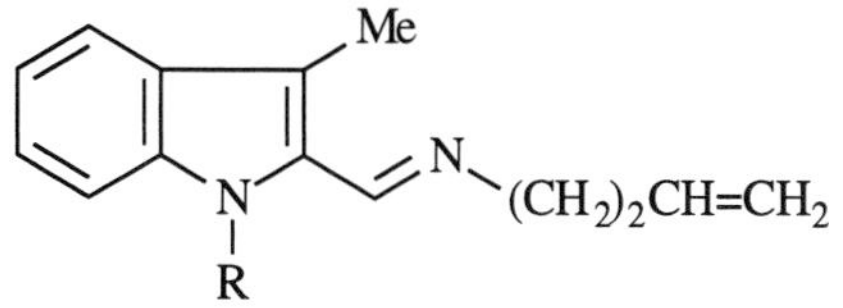

1 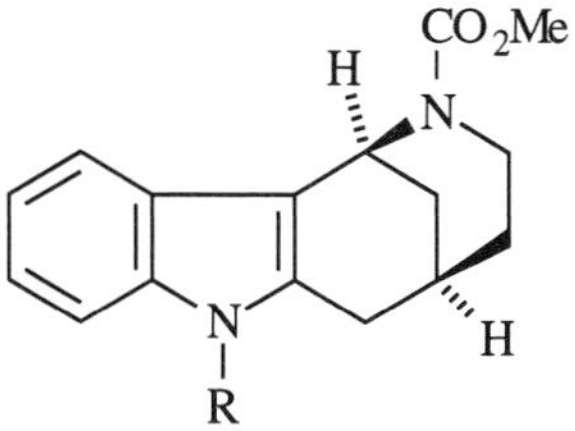 **2**

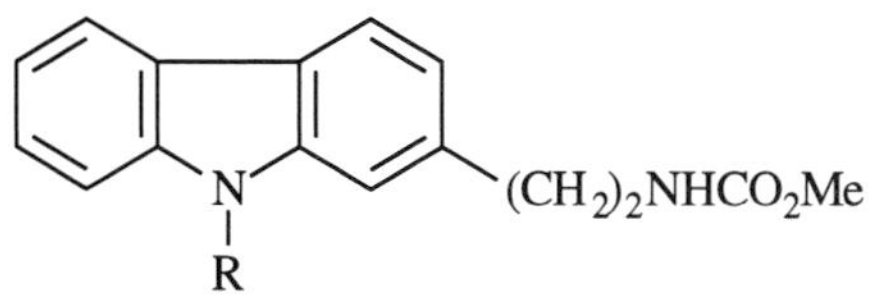

3

$R = 4\text{-MeOC}_6\text{H}_4\text{SO}_2$

199.   A New Synthesis of Substituted Ninhydrins

Ninhydrin is the most commonly employed reagent for revealing latent fingerprints on paper and other porous surfaces, where it reacts with the amino acid deposits released by eccrine sweat glands. Much investigation has been undertaken recently into the preparation of substituted ninhydrins for such applications, in attempts to improve contrast and visualisation (e.g. on banknotes), and a very versatile synthetic route involves heating indenones such as **1** in DMSO. This gives good yields of the ninhydrins.

Suggest a simple synthesis of **1** and a mechanism for the oxidation of **1** to the ninhydrin.

1

200.   Thermal Elimination Reactions Often Fail

In an attempt to prepare the stilbene (**1**), the dihydropyridine (**2**) was pyrolysed at 180°C at 0.1mm. The products obtained, however, were *m*-chlorostyrene (82%) and 2,4,6-triphenyl-pyridine.

Suggest a mechanism for the formation of these products.

1   **2**

201. The Wrong Choice of Reaction Conditions? How not to Prepare an Acid Chloride

Attempts to carry out what are apparently trivial reactions can often lead to completely unexpected results. There are many causes of such anomalous reactions, but perhaps the most common is an injudicious choice of reaction conditions, as illustrated by the following example.

The indole derivative **1** was heated under reflux in thionyl chloride for 5 hours, then excess of thionyl chloride was removed by distillation and the residue was treated with methanol. This did not, however, give the expected product, *viz.* the methyl ester of **1**; instead, the product (63%) was shown to be the tricyclic compound **2**.

Suggest a mechanism for the formation of **2** from **1** under these conditions.

202. Quantitative Yield Isomerisation of a Xylenol Derivative....

A solution of 2-acetoxy-4-methylthio-3,5-xylenol **1** in benzene containing excess triethylamine and methyl isothiocyanate was heated under reflux for 18 hours in an attempt to prepare the *N*-methylthionocarbamate from **1**. The product, however, which was formed in quantitative yield, was found to be 4-[(acetoxymethyl)thio]-3,5-xylenol **2** and it was shown that the same transformation occurred just as readily in the absence of the methyl isothiocyanate.

Give a mechanistic explanation for the formation of **2** from **1**.

203. .... and an Alternative Route to the Starting Material for Problem 202

Reaction of 4-methylsulphinyl-3,5-xylenol **1** in a 2:1 mixture of acetic anhydride-acetic acid at 100°C goes to completion within an hour and gives two products. The major product is 2-acetoxy-4-methylthio-3,5-xylenol **2** and the minor a crystalline solid **A**, (also) $C_{11}H_{14}O_3S$. The ir spectrum of **A** shows bands at 1745, 1660 and 1620 cm^{-1} while the ^{1}H nmr spectrum has bands at δ6.20 (2H, s), 2.18 (3H, s), 2.04 (6H, s) and 1.84 (3H, s). **A** is converted quantitatively into **2** when a solution in benzene is heated under reflux.

Deduce the structure of **A** and give reasonable mechanisms for the formation of **2** and **A** from **1** and for the transformation of **A** into **2**, which was obtained in 50% yield.

204. Hydride-Induced Rearrangements with Indole
Alkaloid Intermediates

During studies on the total synthesis of *Aspidosperma* type alkaloids, unexpected difficulty was encountered in attempts to reduce the amide carbonyl group of the intermediate **1**. Thus, many attempts to reduce **1** with lithium aluminium hydride resulted in reduction of both the amide carbonyl group and the C=C double bond. In an effort to circumvent this problem **1** was reacted with hot phosphorus oxychloride and the intermediate thus obtained treated with sodium borohydride in anhydrous methanol. The product which was isolated, however, was the pentacyclic compound **2**, which was obtained in 50% yield.

Suggest a mechanism for this transformation.

R = COOEt

205. An Unusual "Hydrolysis" Product of 2-Nitrosopyridine

2-Nitrosopyridine is a pale yellow coloured solid, but when it is dissolved in either water or organic solvents the solutions are green in colour. During an investigation of the chemistry of nitrosoazines, it was found that when a suspension of powdered 2-nitrosopyridine in water was stirred at room temperature the solid dissolved gradually and the colour of the solution changed from pale green to light yellow. After 27-30 hours, all of the nitrosopyridine had been consumed and the product (66%) was shown to be 1-(2-pyridyl)-2(1*H*)-pyridone **1**. Closer examination of the reaction showed that (a) there was no conversion of the nitrosopyridine into **1** when 25% dioxane/aqueous phosphate buffer (pH 7.5-8.0) was used, even after 7 days at room temperature; (b) the rate of formation of **1** in water increased with time; and (c) the formation of **1** was acid catalysed: addition of one drop of concentrated sulfuric acid to a 25% dioxane/water solution of the nitrosopyridine resulted in complete conversion into **1** within 2.5 hours. In a standard crossover experiment, stirring of an aqueous acidic solution of a mixture of 2(1*H*)-pyridone and 4-methyl-2-nitrosopyridine gave the pyridylpyridone **2** as the only "dimeric" product.

Give a mechanism to account for the formation of **1**.

1, R = H
2, R = Me

206. A Quinolizine to Indolizine Transformation

Treatment of quinolizinium bromide with two equivalents of piperidine gives a high yield of a product **A**, $C_{14}H_{18}N_2$. Reaction of **A** with phenacyl bromide followed by quenching of the reaction mixture with water gives a product **B** which was originally claimed to be 3-benzoyl-2-vinylindolizine. Subsequent reinvestigation of the structure of **B**, however, showed that it was in fact 3-(2-phenyl-1-indolizinyl)prop-2-ene-1-al **1**.

Deduce the structure of **A** and give a reasonable explanation for the conversion of **A** into **B** ($\equiv$**1**).

1

207. S_NHetAr Reactions Often Proceed with Complications

Treatment of 2-bromo-6-phenoxypyridine with potassium amide in liquid ammonia gave a mixture of three isomeric compounds which were identified as 2-amino-6-phenoxypyridine (28%), 4-amino-2-phenoxypyridine (9%), and 2-methyl-4-phenoxypyrimidine (50%).

Suggest mechanisms for the formation of these three products.

208. Rearrangement During Recrystallisation

Condensation of the pyrrolidine enamine of cyclohexanone with 1,1-dicyano-2,2-dimethylcyclopropane proceeds smoothly in refluxing dry xylene and gives the expected adduct in 76% yield. Recrystallisation of the adduct from 95% ethanol, however, gave a 91% yield of a product which no longer contained the pyrrolidine group but whose spectral data clearly showed the presence of a ketone group and an enaminonitrile function. Hydrolysis of this latter product with phosphoric acid/acetic acid gave 5-(2-oxo-4,4-dimethylcyclopentyl)pentanoic acid in 83% yield.

Elucidate the reaction sequence and give mechanisms for the transformations.

209.   Another Attempt to Reduce a Ketone Goes Wrong

When the amino ketone **1** was heated at 150-160°C for 15 minutes with the constant boiling liquid salt "trimethylammonium formate" (TMAF, 5HCOOH.2NMe₃), two products were obtained, the major of which (58%) was shown to be **2**.

Give a mechanism for the formation of **2**.

1   **2**

210. A More Complex Benzocyclobutane To Isochroman-3-one Rearrangement (c.f. Problem 138)

Heating of the benzocyclobutene derivative **1** in degassed *o*-dichlorobenzene at 180°C for 2 hours gives the spirocyclic product **2** in 88% yield.

Give a mechanism for this transformation.

1	**2**

211. Indoles by Solvometalation Ring Closure

In an interesting approach to the mitosenes, Danishefsky showed that treatment of **1** with mercuric acetate in THF containing sodium bicarbonate leads directly to the indole derivative **2** in over 50% yield.

Suggest a mechanism for this transformation.

1	**2**

$$R^1 = CH_2Ph$$
$$R^2 = TBDMS$$

212. An Unusual - but Inefficient - Synthesis of Methyl 1-Hydroxynaphthalene-2-carboxylate

Methyl 1-hydroxynaphthalene-2-carboxylate is formed in 19% yield when homophthalic anhydride and methyl propiolate are heated in toluene at 150° for 24 hours.

Suggest at least two mechanisms for the formation of the naphthalene ester.

213. Synthesis of α-Arylalkanoic Acids from Acetophenones

An elegant, if rather long, procedure for the synthesis of α-arylalkanoic acids is exemplified by the conversion of anisole into methyl α-(4-methoxyphenyl)propionate by the following sequences of steps:

1.   Anisole + $MeCH_2COCl/AlCl_3 \rightarrow$ **A**
2.   **A** + Br_2/dioxane $\rightarrow$ **B**
3.   **B** + MeONa/MeOH $\rightarrow$ **C**
4.   **C** + TsCl/pyridine $\rightarrow$ **D**
5.   **D** heat in aqueous methanol in the presence of $CaCO_3 \rightarrow$ product.

Elucidate the overall scheme and give mechanisms for steps 3 and 5.

214. Triazene-Triazole-Triazole Interconversions

Recrystallisation of 3-(cyanomethyl)-1-(*p*-nitrophenyl)triazene from absolute ethanol gives 5-amino-1-(*p*-nitrophenyl)-1,2,3-triazole in 80% yield. When either of these compounds is heated under reflux in ethanol for an hour, however, they are smoothly isomerised into 4-(*p*-nitrophenylamino)-1,2,3-triazole in 99% yield.

Give mechanisms to explain these observations.

215. Reissert Compounds as Precursors to Novel Phthalides

Heating of the Reissert compound **1** with sodium hydroxide in aqueous methanol gives the phthalideisoquinoline **2** in 58% yield.

Suggest a mechanism for this transformation.

| 1 | 2 |

216. Di-t-butylacetylene does not Cycloadd to 2-Pyrone

Condensation of 2-pyrone with dimethyl acetylenedicarboxylate and diethylacetylene proceeds in the expected manner. There is, however, no reaction between di-t-butylacetylene and 2-pyrone, even after solutions of the reactants in bromobenzene have been heated in a sealed tube at 210°C for 5 days. Instead, *trans*-cinnamic acid is obtained in low yield.

Give a mechanism for formation of the cinnamic acid.

217. A Failed Thorpe-Dieckmann Cyclisation : "Obvious" Reactions are *not* Always Well Behaved . . . . . .

During synthetic studies on the benzophenanthridine alkaloids, the nitrile ester **1** was treated with sodium hydride in THF in the expectation that a Thorpe-Dieckmann type of reaction would occur. The product was found, however, to be the naphthopyran derivative **2**, which was obtained in 44% yield.

Give a mechanism for the formation of **2**.

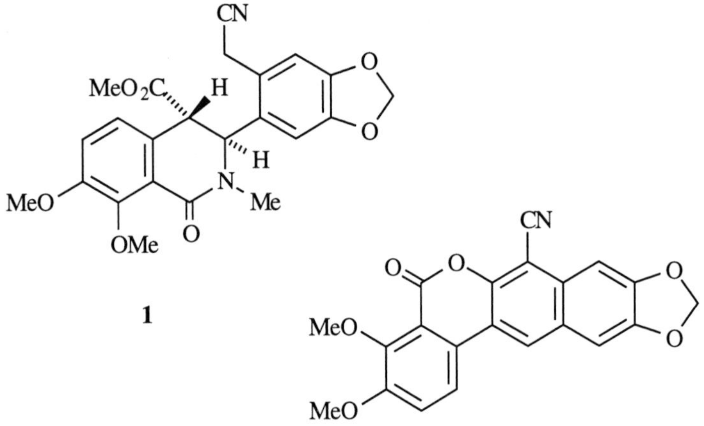

1

2

---

218. . . . . . . and can lead to Remarkable Rearrangements : A Failed Thorpe-Ziegler Cyclisation

Japanese workers recently designed a synthesis of 5-amino-2,3-dihydrothiepino[2,3-*b*]pyridine-4-carbonitrile **2** based on Thorpe-Ziegler cyclisation of 2-(3-cyanopropylthio)pyridine-3-carbonitrile **1**. Treatment of **1** with potassium t-butoxide however, did not, give **2**, but produced the thieno[2,3-*h*][1,6]naphthyridine **3** in 82% yield.

Suggest a mechanism for the transformation **1** → **3**.

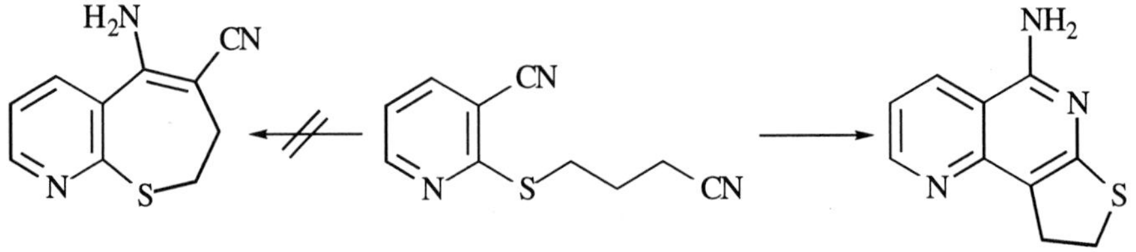

2   1   3

219.   A Remarkably Stable Tertiary Alcohol by Solvolysis of a Primary Tosylate

Solvolysis of nopol tosylate **1** in aqueous ethanol was reported to give a mixture of **2** (31%), **3** (6%) and a tertiary alcohol (41%) which was assigned structure **4**. The tertiary alcohol was remarkably stable to acid and was best purified by refluxing in strong sulfuric acid. This decomposed all other organic products but left the tertiary alcohol unchanged. Subsequently, it was found that solvolysis of **1** in acetic acid containing sodium acetate gave the acetate of the tertiary alcohol as virtually the exclusive product in 91% yield, and the structure of the alcohol was shown to be **5**, not **4**.

Give a mechanistic rationale for the conversion of **1** into **5**.

220.   "Anionic Activation" for the Preparation of Fluoroheterocycles

The following is a typical example of a process described by Russian workers as "anionic activation", and used for the rapid construction of fluoro-substituted heterocycles: Treatment of 2-(trifluoromethyl)aniline with either lithium phenylacetylide or the lithium enolate of acetophenone in THF at -60°C gave 4-fluoro-2-phenylquinoline in 25-40% yield.

Suggest a mechanism for this transformation and predict the structure of the product that would be formed if the same concept were applied to the condensation of 2-(trifluoromethyl)benzyl chloride with 2-benzo[*b*]thienyllithium.

221. Synthesis of a 5-(2-Quinolyl)pyrimidine

Elucidate the following sequence of reactions and give a reasonable mechanism for the formation of intermediate **C**:

1. Condensation of *m*-chloroaniline with ethyl acetoacetate in the presence of a catalytic amount of HCl to give **A**.
2. Heating of A in refluxing diphenyl ether to give **B**, $C_{10}H_8ClNO$.
3. Treatment of **B** with $POCl_3/DMF$ at 100°C followed by quench with ice and dilute NaOH to give **C**, $C_{14}H_{12}Cl_2N_2O$.
4. Reaction of **C** with guanidine in ethanol at reflux to give **D**, $C_{13}H_8Cl_2N_4$.

222. "Obvious" and "Non-Obvious" Pathways to a Highly Substituted Pyridine and Aniline

Condensation of (*E*)-4-phenylbut-3-en-2-one (**1**) with malononitrile (**2**) in boiling sodium methoxide/methanol was found to give a mixture of 3-cyano-2-methoxy-6-methyl-4-phenylpyridine (**3**) and 2,6-dicyano-3-methyl-5-phenylaniline (**4**). Simple and "obvious" mechanisms were advanced to account for the formation of (**3**) and (**4**).

A more recent reinvestigation of this reaction, however, has shown that the rate of formation of both (**3**) and (**4**) varies with experimental conditions. In particular, the relative stoichiometry of (**1**) and (**2**) was found to be important, and the aniline (**4**) could be obtained predominantly at a ratio of (**2**):(**1**) of 2:1. The investigators then suggested a "non-obvious" pathway for the formation of (**4**).

Give both the "obvious" pathways for the formation of (**3**) and (**4**) and the "non-obvious" pathway for the formation of (**4**).

223. A 1-Isoquinolone Synthesis

Reaction of methyl isocyanoacetate with methyl 2-formylbenzoate in the presence of sodium hydride in DMF at 30-40°C gives methyl 1-oxo-1,2-dihydroisoquinoline-3-carboxylate in 42% yield.

Give a mechanism for this transformation.

224. Selective Cleavage of the Mycinose Sugar from the Macrolide Antibiotic Tylosin

Mild acid treatment of the complex 16-membered ring macrolide antibiotic tylosin **1** results in selective hydrolysis of the mycarose sugar; hydrolysis of both the mycarose and mycinose sugars occurs under more vigorous acid conditions. A procedure for selective cleavage of the mycinose sugar was developed by chemists from Pfizer Central Research which involved treatment of the derivative **2** with a primary amine hydrochloride in warm isopropanol. When benzylamine hydrochloride was used the exclusive products were the macrolide residue formed by cleavage of the mycinosyl glycosidic residue and *N*-benzyl-2-acetyl-3-methoxylpyrrole.

Give a mechanistic explanation for glycoside cleavage and formation of the pyrrole derivative.

1

2, M = suitably protected macrolide

225. Unexpected Formation of an Enamide

The pyrrolidine derivative **1** was treated with trifluoroacetic anhydride in the expectation that intramolecular cyclisation would result. The only products isolated, however, were the isomeric enamines **2a** and **b**.

Give a mechanism for the transformation of **1** into **2a, b**.

1

2a, $R^1 = 3\text{-MeOC}_6H_4$
$R^2 = \text{COCF}_3$
b, $R^1 = \text{COCF}_3$,
$R^2 = 3\text{-MeOC}_6H_4$

226. Highly Functionalised Furans from 3-Bromochromone

3-Bromochromone reacts readily at room temperature with acetylacetone in chloroform in the presence of two equivalents of DBU. The product, isolated in 83% yield in the usual manner after quenching of the reaction mixture with 5% HCl, is 3-acetyl-5-(2-hydroxybenzoyl)-2-methylfuran. A variety of β-diketones and β-keto esters react similarly to give moderate to excellent yields of highly functionalised trisubstituted furans.

Explain this novel synthesis in mechanistic terms.

227. A One-Pot Benzene → Naphthalene Transformation

Annulation of a benzene derivative to the naphthalene framework normally requires a number of steps. A recent one-pot procedure involves heating the benzhydrols **1** in toluene with a five-fold excess of maleic anhydride for several hours, and gives the 1-arylnaphthalene derivatives **2** in moderate yield.

Give a mechanism for this transformation.

1   **2**

228. Vicarious Nucleophilic Substitution Routes from Simple to Complex Phenols

VNS in recent years has become a valuable and in many instances powerful method for the regiospecific functionalisation of aromatic and heteroaromatic compounds. In a recent study, for example, it was shown that aluminium chloride-catalysed phenylsulfinylation of *p*-cresol followed by treatment of the product with (a) thionyl chloride, or (b) acetic anhydride/sodium acetate, or (c) *p*-cresol and TFAA gave, respectively, (a) 2-chloro-4-methyl-6-(phenylthio)phenol, or (b) 2-acetoxy-4-methyl-6-(phenylthio)phenol, or (c) 2,2'-dihydroxy-5,5'-dimethyl-3-(phenylthio)biphenyl. Yields varied from moderate to good (30-78%).

Give a mechanistic explanation for these transformations.

229. Lewis Acid Catalysed Rearrangement of Humulene 8,9-Epoxide

The three mono-epoxides of humulene **1** are naturally occurring, and it is believed that they are *in vivo* precursors of other bicyclic and tricyclic sesquiterpenes. *In vitro* experiments have demonstrated that the 1,2- and 4,5-epoxides undergo facile acid-catalysed rearrangement, and it has been shown recently that treatment of a chloroform solution of the 8,9-epoxide with tin(IV) chloride at -60°C for 15 minutes gives a variety of hydrocarbons and one major product (25%), the alcohol **2**.

Suggest a mechanism for the formation of **2**.

230. From a Dihydrofuran to an Indole-3-acetate

Treatment of ethyl *N*-2-iodophenylcarbamate with 2,5-dihydro-2,5-dimethoxyfuran in DMF in the presence of Hünig's base, TEBA and a catalytic amount of palladium(II) acetate at 80°C for 10 hours gave the expected 3-aryl-2,5-dimethoxy-2,3-dihydrofuran **1** as a mixture of diastereomers. Crude **1** was stirred at room temperature in DCM containing 6% v/v TFA to give methyl 1-ethoxycarbonylindole-3-acetate **2** in 65% overall yield.

Elucidate the reaction sequence and suggest a mechanism for the conversion of **1** into **2**.

231. Reaction of Fervenulin 4-Oxide with DMAD : The Role of Solvent

Treatment of fervenulin 4-oxide **1** with 1.5 equivalents of DMAD in toluene at 95°C for 3 hours resulted in the formation of the pyrrolopyrimidine **2** in 62% yield. It was suspected that water in the solvent was participating in the formation of **2**. Evidence in support of this hypothesis was obtained by carrying out the reaction in anhydrous toluene. The product obtained (56%) was the pyrrolopyrimidine **3**. Unexpectedly, use of ethanol as the solvent gave the pyrrolopyrimidine **4** in 34% yield, together with 29% of recovered starting material.

Suggest mechanisms to account for the formation of **2, 3** and **4** under the different conditions.

1

2

3

4

232. Rearrangement of 4-Quinazolinylhydrazines

N-(2-Hydroxyethyl)-*N*-(6,7,8-trimethoxyquinazolin-4-yl)hydrazine **1** undergoes smooth rearrangement to the isomer **2** at near neutrality (pH 4-7.7) in aqueous solution, but is hydrolysed to quinazolin-4-one **3** at higher and lower pH. Chemists at Zeneca have studied these processes in detail and shown that they are remarkably complex.

Suggest mechanisms for the formation of **2** and **3** from **1**.

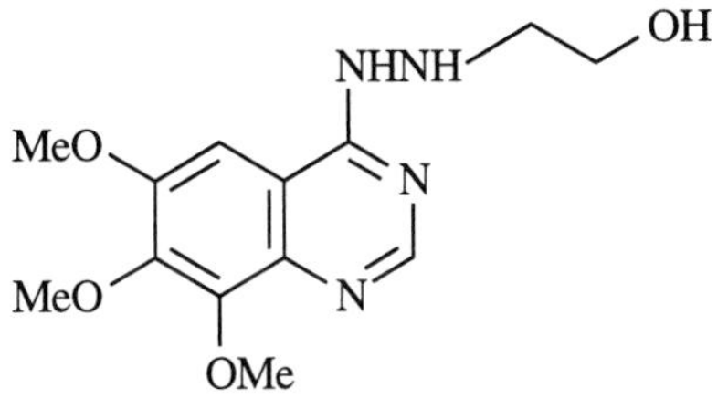

2

233. A Failed Approach to the Oxetan-3-one System

In a recent approach to the preparation of dihydro-3(2*H*)-furanones by chemists from Thailand it was shown that consecutive, one-pot conventional treatment of 1,3-dithiane in THF with equimolar *n*-butyllithium, an epoxide, *n*-butyllithium, an aldehyde or ketone and, finally, methanesulfonyl chloride gave a spirocyclic dithioacetal in good (*ca.* 70%) yield. Hydrolysis with HgO/HgCl$_2$ gave the corresponding dihydrofuranone.

An attempt to extend this approach to the preparation of the oxetan-3-one system, *i.e.* use of an aldehyde in place of an epoxide, failed. Instead, a 2-acyl-3-substituted-1,4-dithiepane derivative was obtained.

Elucidate the above reaction sequences and give a mechanism for formation of the ring expanded product.

234.   Tricyclics from Furfural

Elucidate the synthesis of **1** from furfural, and identify X:

Furfural $\xrightarrow{\text{1 - 4}}$

1

1.   Propargylamine/catalytic PPTS followed by NaBH$_4$
2.   Di-t-butyl dicarbonate/catalytic DMAP/TEA
3.   KO^{t}Bu (5 equiv)/^{t}BuOH
4.   DMAD in the presence of PPTS/TMOF

235.   An Ylide-Based Synthesis of 4-Phenylisocoumarin

Corey's procedures for the conversion of ketones into epoxides using sulfonium or oxosulfonium ylides have found widespread use in organic synthesis. An attempt to apply the method to methyl 2-benzoylbenzoate, however, gave 4-phenylisocoumarin in 52% yield when dimethyloxosulfonium methylide was used.

Suggest an explanation for this transformation.

It was found during this study that it was important to use only one equivalent of the methylide for preparation of the isocoumarin. When two equivalents were used, none of the isocoumarin was formed but, instead, a stable "ylide" was obtained in 96% yield.

Give a structure for the "ylide" and a mechanism for its formation.

236.   Model Studies for Mitomycin A Synthesis Lead to a New Preparation of Pyrroloindoles

Mitomycin A, **1**, is a natural product which shows potent antibiotic and antitumor activity, and has been the subject of extensive synthetic studies. French workers have described a simple procedure for one-pot preparation of the tricyclic skeleton of the mitomycins. Thus, simply mixing and stirring a solution of nitrosobenzene and *(E,E)*-hexa-2,4-dienal in absolute ethanol at room temperature for two days gave the pyrroloindole derivative **2** in about 50% yield together with the betaine **3** (35%) and the pyrrole **4** (1%).

Give a mechanistic explanation for the formation of each of the products **2-4**.

1	**2**

3	**4**

237. Formation of *N*-Cyanofluoren-9-imine from 9-Dinitromethylenefluorene

Reaction of 9-dinitromethylenefluorene (DNF) with nucleophiles such as CN⁻, MeO⁻ and PhCH2S⁻ proceeds smoothly to give the corresponding stable carbanions in quantitative yield. When DNF is treated with a solution of sodium azide in DMSO at ambient temperature, however, there is a rapid reaction which is complete within two minutes and which is accompanied by the vigorous evolution of gas. The product is *N*-cyanofluoren-9-imine.

Give a mechanistic explanation for formation of the *N*-cyanofluoren-9-imine.

238. Methylphenylacetic Acids from Butenolides

2-Methylphenylacetic acid is obtained in 46% yield when the butenolide **1** is heated with five equivalents of molten pyridine hydrochloride at 220°C for three hours. Reaction of the butenolide **2** under precisely the same conditions, however, gives 3-methylphenylacetic acid as the only aromatic product in 52% yield.

Give mechanistic explanations for these observations.

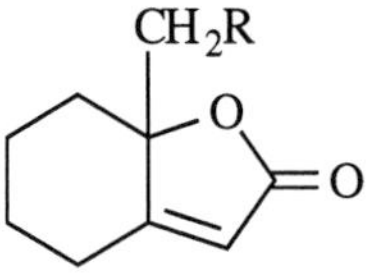

1, R = OH
2, R = NMe$_2$

239. Base-Induced Quinoline Rearrangements

Compound **2** is a potent cyclic AMP phosphodiesterase inhibitor, and can be prepared in almost quantitative yield by treatment of **1** with 1.1 equivalents of sodium methoxide in refluxing methanol for 15-30 minutes. When either **1** or **2** is refluxed with excess of sodium methoxide in methanol, however, the quinoline derivative **3** is obtained in 92% yield.

Suggest a mechanism for the formation of **3**.

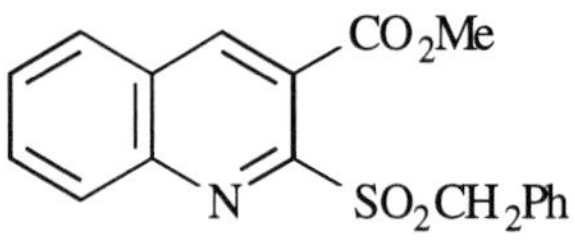

1

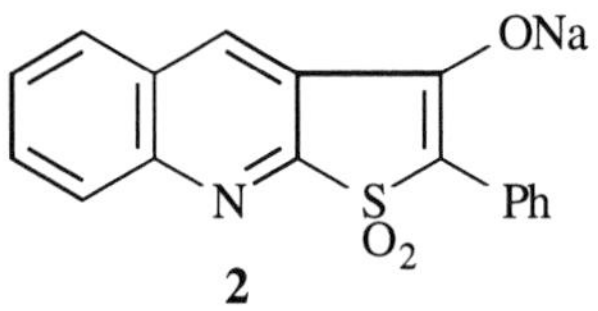

2

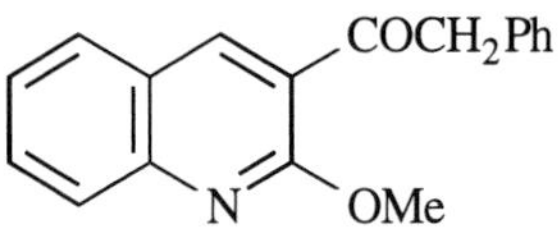

3

240. The Angucyclines : Rearrangement of Angular to Linear Tetracycles

Many of the polycyclic antibiotics are based on linear assemblies of three or more six-membered rings, such as the tetracycline and anthracycline antibiotics (examples are chlortetracycline and daunorubicin). Since 1966, however, more that 100 secondary metabolites have been isolated which have an unsymmetrically (angularly) assembled tetracyclic ring frame. These antibiotics are referred to as the "angucycline" group, and aquayamycin **1** is typical. Many of the angucycline antibiotics rearrange to linear tetracyclic systems extremely easily. **1**, for example, is rearranged to three different linear tetracycles on treatment with either base or acid, or on u.v. irradiation. Simply heating **1** briefly at 220°C gives **2**.

Suggest a mechanism for the formation of **2** from **1**.

1   **2**

241. 1,2-Dihydropyridines from 2,3-Dihydro-4-pyridones

1-Acyl-2-alkyl-2,3-dihydro-4(1*H*)-pyridones **A** are readily available heterocycles in both racemic and chiral form. 1-Acyl-2-alkyl-1,2-dihydropyridines **B** are much less readily accessible, especially enantiopure, but are much sought after building blocks for alkaloid synthesis. A very efficient (83-96%) and simple procedure has now been developed for the **A** → **B** transformation, illustrated as follows: treatment of 1-allyloxycarbonyl-2-cyclohexyl-2,3-dihydro-4-pyridone with one equivalent of the Vilsmeier reagent in trichloroethylene at room temperature gave 1-allyloxycarbonyl-4-chloro-2-cyclohexyl-1,2-dihydropyridine in 92% yield.

Suggest a mechanism for this transformation.

242.   The Vitamin K to Vitamin K Oxide Transformation

Much attention has focused on vitamin K **1** because of its function as an obligatory cofactor in enzymic sequences central to blood clotting. The role of molecular oxygen in the formation of vitamin K oxide **2** has been studied intensively, and the mechanism of the **1** → **2** transformation has been the subject of much controversy. Oxidation of **1** with basic hydrogen peroxide also gives **2**, and two obvious mechanisms can be postulated for this model oxidation. ^{18}O labelling studies have been used to distinguish between these mechanisms.

What are the "obvious" mechanisms for oxidation of **1** to **2**? How does the use of $H_2^{18}O_2$ enable a choice to be made between them?

243.   Rearrangement During Intramolecular Cyclisation to the Indole 4-Position

Intramolecular Friedel-Crafts cyclisation of the acid **1** to the 4-position takes place perfectly normally to give the tricyclic ketone.   Under almost identical conditions ($AlCl_3$/nitrobenzene/0°C), the acid chloride derived from **2** cyclised smoothly to give a 50% yield of a tricyclic ketone. This however, was shown to be **3**, and *not* the expected product.

Suggest a mechanistic explanation for the conversion of **2** into **3**.

244. 4-Quinolone Antibacterials : A New Synthesis

Chemists at Pfizer have developed short routes to novel tricyclic quinolone derivatives of interest as potential antibacterials. The overall transformation is illustrated by **1** → **2**.

i,   PhCl, reflux (85%)
ii, NBS, CCl$_4$ (58%)
iii, KOH, EtOH, THF (47%)

Elucidate the overall sequence, giving mechanisms.

Subsequent exploitation of the mechanistic aspects of the **1** → **2** transformation enabled the Pfizer chemists to develop a one step synthesis of **2** in 54% yield from a starting material similar to **1** and just as accessible. What is the one step synthesis?

245. An Efficient Synthesis of Fused 1,3-Dithiol-2-ones

A very elegant, simple, versatile and high yielding synthesis of fused 1,3-dithiol-2-ones has been described recently. In a representative example, treatment of 4-benzoyloxy-1-bromo-2-butyne with potassium *O*-methyl xanthate in methanol gave the expected xanthate ester in excellent yield. Refluxing of this ester in chlorobenzene for one hour in the presence of dimethyl fumarate gave **1** in quantitative yield.

Elucidate the reaction sequence and give a mechanism for the formation of **1**.

246.   The "Double Functional Group Transformation" : Terminally Unsaturated Nitriles from 1-Nitrocycloalkenes

There is much current interest in the design and development of the "double functional group transformation" concept, an excellent illustration of which is the now well known and widely exploited Eschenmoser fragmentation reaction. A recent example of the "double functional group transformation" is the one-flask conversion of 1-nitrocycloalkenes into terminally unsaturated nitriles by treatment first with trimethylsilylmethylmagnesium chloride (1.8 eq.) in THF at -20°C and then, *in situ*, with PCl$_3$ (2.5 eq.) at 67°C. 2-Nitrobicyclo[2.2.1]hept-2-ene, for example, gave *cis*-1-cyano-3-vinylcyclopentane in 33% yield, and similar yields of ene-nitriles were obtained from a variety of monocyclic and bicyclic 1-nitrocycloalkenes.

Suggest a mechanism for this "double functional group transformation".

247.   Lewis Acid-Catalysed Condensation of Indole with 1,3-Cyclohexanedione

Addition of freshly distilled BF$_3$.OEt$_2$ to a solution of indole and 1,3-cyclohexanedione in methylene chloride at 0°C gave a mixture of **1** and **2** in 55 and 20% yield respectively.

Suggest mechanisms for these transformations.

1   **2**

248.   Serendipitous Preparation of a Pyrrole Precursor to Porphyrins

Ethyl 5-methylpyrrole-2-carboxylate has been used as a precursor to porphyrins, and is accessible by a number of multistage syntheses. A new, one-pot preparation has been discovered recently by accident, which gives this pyrrole in 35-40% yield from cheap and readily available starting materials.

American workers needed to prepare the bis-amino acid **1** and adopted a literature procedure in which two equivalents of diethyl acetamidomalonate were to be alkylated with one equivalent of 1,4-dichloro-2-butyne using two equivalents of sodium ethoxide in hot ethanol. Hydrolysis and decarboxylation of the dialkylated malonate would then give **1**. This alkylation reaction was carried out, but ten equivalents of sodium ethoxide were used rather than two. This resulted in formation of ethyl 5-methylpyrrole-2-carboxylate in *ca.* 40% yield. Further study showed that the reaction to produce the pyrrole required equimolar amounts of the acetamidomalonate and the dichlorobutyne, excess of sodium ethoxide, and heating. No pyrrole was formed at room temperature.

Devise a mechanism to account for the formation of ethyl 5-methylpyrrole-2-carboxylate.

1

249.   A Chromone Ring Contraction

Nitration of chromone-3-carbaldehyde using fuming nitric acid and concentrated sulfuric acid at ice-bath temperature gives 6-nitrochromone-3-carbaldehyde. When the same reaction is carried out using red fuming nitric acid in place of fuming nitric acid, the product (48% yield) is 5-nitro-2,3-benzofurandione (Z)-2-oxime. Reaction of chromone-3-carboxylic acid under the same conditions gives the same oxime.

Suggest an explanation for the formation of the oxime.

250. A Simple Conversion of Hydroquinone to a Benzofuran, but Which Mechanism?

Reaction of one equivalent of hydroquinone with 0.5 equivalent of dimethyl chloromalonate in methanol containing one equivalent of sodium methoxide proceeds exothermically and with the development of a transient dark green colour to give sodium 5-hydroxy-3-methoxycarbonyl-benzo[*b*]furan-2-olate in 65% yield.

Several mechanisms can be suggested to account for this transformation. Identify the most likely mechanism.

251. A Remarkable Loss of One Carbon Atom in the Indole Alkaloid Field

The product from the reaction of anhydrovinblastine N_b-oxide **1** with trifluoroacetic anhydride in methylene chloride was treated, after evaporation of the excess of the reagent, with a mixture of water and THF. This gave a mixture of products, only one of which could be obtained pure (27%), namely 5'-noranhydrovinblastine **2**.

Suggest a reasonable mechanism for the transformation of **1** into **2**.

252. A "Non-Obvious" Cycloaddition Reaction

The diene **1** reacts exothermically with dimethyl acetylenedicarboxylate in benzene at room temperature to give dimethyl 4-*N*,*N*-dimethylamino-5-methylphthalate in 70% yield.

Suggest a mechanism for this transformation.

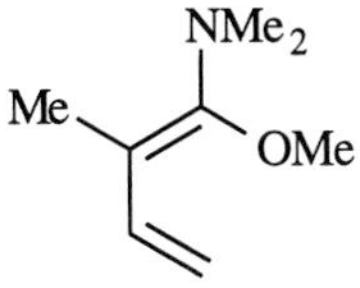

1

253. More Radical Cascades and a Formal [2+2+2] Cycloaddition of a Dienyne

The dienyne **1** (0.77 mmol) was treated with triphenyltin hydride (1.1 mmol) and triethylborane (0.38 mmol) in toluene (0.016 M) at room temperature for 3-6 hours. Removal of the solvent and flash chromatography of the residue gave **2** as the major product (54%), together with **3** (15%) and **4** (12%). All three products were obtained diastereomerically pure.

Suggest mechanisms for the formation of **2**, **3** and **4**.

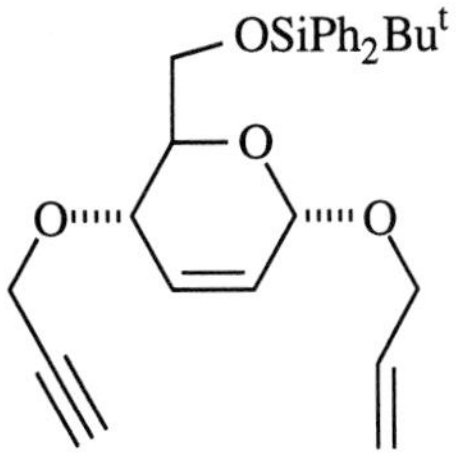

1 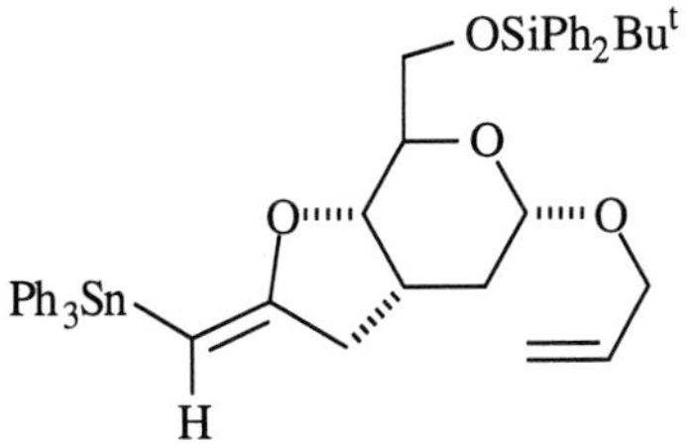   **2**

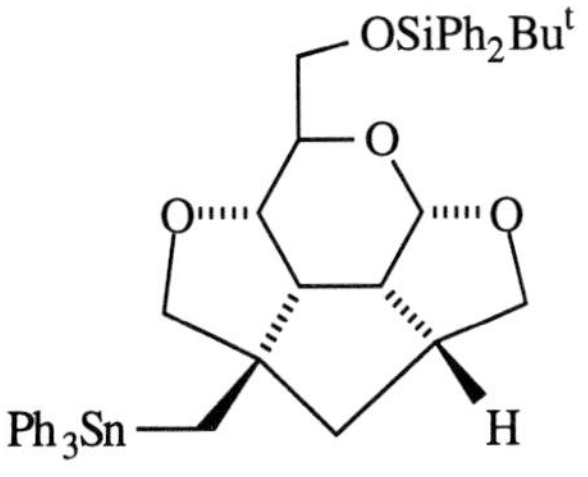

3 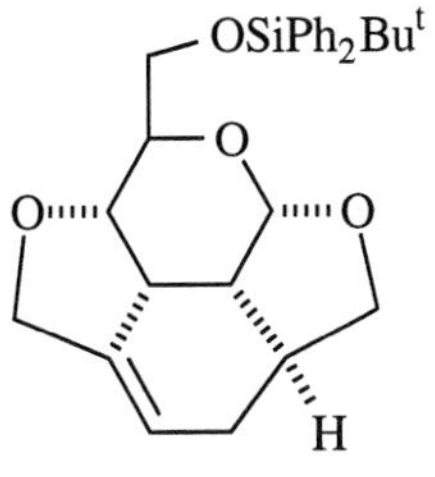   **4**

254. LTA-Induced Acetoxylation and Rearrangement of a Phenol

Oxidation of the phenol **1** with LTA in acetic acid gave two products, the major of which (60%) was shown to be the phenol **2**.

Suggest a mechanism for this transformation.

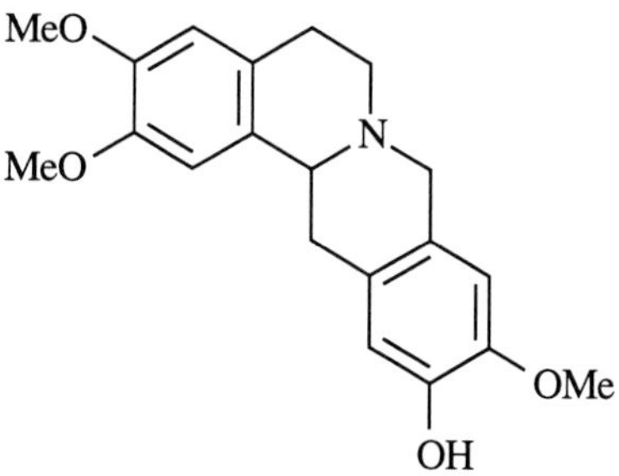

1

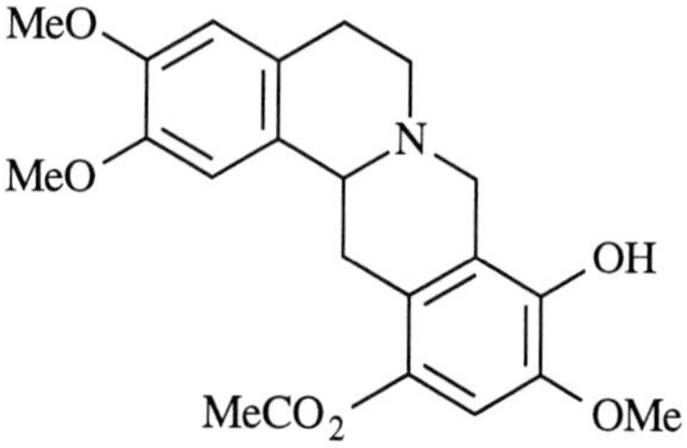

2

255. How Cocaine Decomposes at 550°C

Flash vacuum pyrolysis of cocaine **1** at 550°C results in complete conversion to thermal degradation products. The major products isolated were benzoic acid (100%), *N*-methylpyrrole (74%) and methyl 3-butenoate (60%).

Give a mechanistic explanation for the formation of these products.

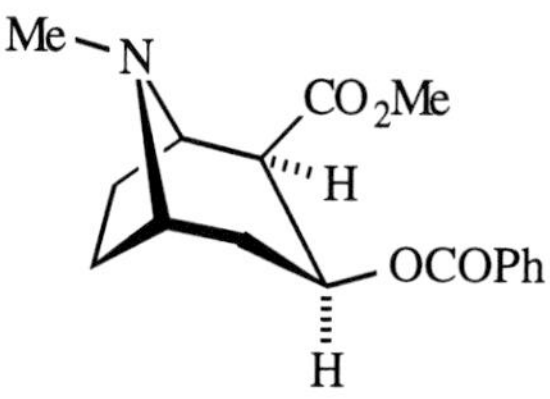

1

256. A Highly Efficient Anilide to Benzimidazole Transformation

Treatment of anilides of the type **1** with a mixture of formic acid and hydrogen peroxide results in direct formation of the benzimidazole derivative **2** in very good yield (85-95%).

Suggest a mechanism for this transformation.

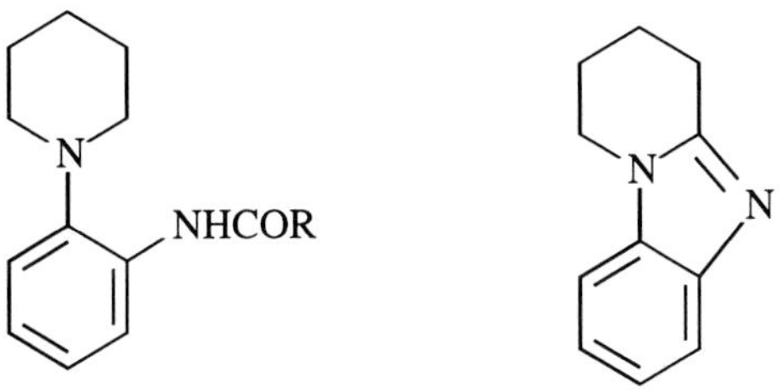

R = alkyl, aryl

1   **2**

257. The Baeyer-Drewson Synthesis of Indigo

The classical Baeyer-Drewson synthesis of indigo **1** consists of reaction of *o*-nitrobenzaldehyde with acetone in the presence of aqueous sodium hydroxide.

Suggest a mechanism for the formation of indigo under these conditions.

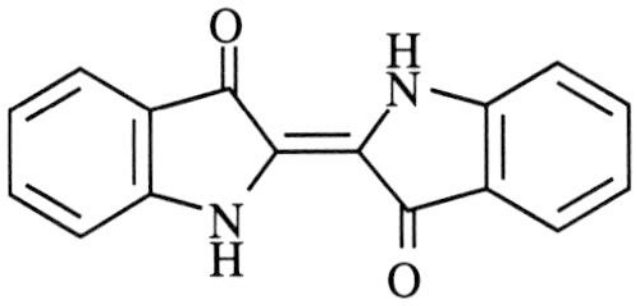

1

258.   The Marschalk Reaction

It was stated in a Tetrahedron Report that the Marschalk reaction is "By far the most important reaction for anthracyclinone synthesis using anthraquinones as starting materials. . . . ."

The general process, which involves reaction of aldehydes with anthrahydroquinones, is outlined in the conversion **1** to **2**.

Suggest a mechanism to account for the Marschalk reaction.

259.   A 1,3-Cyclohexanedione to 2-Cyclohexenone Conversion

The reaction of 2,2-disubstituted 1,3-cyclohexanediones **1** with dimethyl methanephosphonate in THF in the presence of LDA gives 3-substituted 2-cyclohexenones **2** in moderate to very good yields.

Suggest a mechanism for this transformation.

260.   Reaction of a Steroidal Olefin with Br₂/AgOAc

Treatment of the steroid derivative **1** with bromine and silver acetate in a mixture of chloroform and pyridine at low temperature followed by quenching of the reaction mixture with aqueous acid gave a mixture of compounds, one of which was shown to be **2**.

Give a reasonable mechanism for the formation of **2**.

1 R = OAc, R^1 = H
2 R = OH, R^1 = OAc

261. A Most Unusual Synthesis of Tropones from Phenols

American scientists have been studying intramolecular phenolic cyclisations of the type **1** to **2**, and yields are generally good. The transformation **3** to **4**, which proceeds in 80% overall yield, has been described in a review article as "a fantastically short and efficient synthesis of such tropone systems".

Give a mechanism for the conversion of **3** into **4**.

1   **2**

3   **4**

A: (i) KOH; (ii) K$_3$Fe(CN)$_6$; (iii) H$^+$

262.   Side-Chain Manipulation with a Purine Derivative : Unexpected Formation of a Thietane

Reaction of 6-mercaptopurine with sodium bicarbonate and epichlorohydrin in ethanol for 7 days at room temperature gave the expected chlorohydrin derivative. Unexpectedly, when this was treated with two equivalents of sodium methoxide in methanol at room temperature for 24 hours it did not give either epoxide or methoxy containing derivatives. Instead, the product, obtained in 32% yield, was the substituted thietane **1**.

Suggest a mechanism for the formation of **1**

1

263.   An Isoxazoline to Pyridine *N*-Oxide Transformation

Treatment of 5-cyanomethylisoxazolines **1** with a catalytic amount of DBU in boiling xylene gives good to excellent yields of 6-substituted-2-aminopyridine *N*-oxides.

Give a mechanism for this type of transformation.

1

264.   A Simple Synthesis of the Lignan Carpanone

The lignan carpanone **1** would present a formidable synthetic challenge if it were to be constructed from simple starting materials in a linear, stepwise manner. It can, however, be assembled rapidly by the following sequence of four simple reactions (none of the intermediates need be   rigorously purified) : (i) alkylation of 3,4-methylenedioxyphenol with allyl chloride/NaOEt/EtOH proceeded normally to give **A**, $C_{10}H_{10}O_3$, which (ii) underwent smooth isomerisation to **B** on being heated at 170°C for 1 hour. (iii) **B** was isomerised to **C** on treatment with KO^tBu/DMSO at 80°C for 30 minutes, and (iv) treatment of **C** with a mixture of NaOAc/Cu(OAc)$_2$ in aqueous methanol at room temperature for 5 minutes gave carpanone **1**.

Elucidate the reaction sequence and give a mechanistic explanation for the conversion of **C** into carpanone **1**.

1

265. Cyclopentaquinolines by Tandem Reactions

The first "tandem reactions forming geminal carbon-carbon bonds" were reported in 1991, and the following example is representative. A mixture of 5-iodo-1-pentyne (1 eq.) and *p*-fluorophenyl isocyanide (5 eq.) in t-butylbenzene containing hexamethylditin (1.5 eq.) was irradiated under nitrogen in a pyrex flask at 150°C for several hours until all of the alkyne had been consumed (followed by ^{1}H nmr). Standard work up gave a mixture of the cyclopentaquinolines **1** and **2** in 56% yield and 85/15 ratio.

Give a mechanistic explanation for the production of **1** and **2**.

1   **2**

266.   Epimerisation of Herqueinone

The mould metabolite herqueinone **1** undergoes facile epimerisation at C-4 to give *enantio-* isoherqueinone when heated under reflux for one hour with anhydrous potassium carbonate in anhydrous acetone.

Suggest a mechanism for this epimerisation.

1

267.   A New Method for Amide Bond Formation

Possibly the most important condensation reaction is that between a carboxylic acid and an amine to give an amide. A great many methods are known by which this formal dehydration process may be carried out, almost all of which involve the two step sequence : (i) activation of $CO_2H \rightarrow$ COX, where X is a leaving group; and (ii) aminolysis of RCOX. Japanese workers have recently advocated the use of 2,2-dichloro-5-(2-phenylethyl)-4-trimethylsilyl-3-furanone (**1**, "DPTF") for carboxyl activation, and its use for peptide formation is illustrated by the   representative conversion **2** → **3**. The byproduct formed from DPTF in these reactions is 5-(2-phenylethyl)-4-trimethylsilylfuran-2,3-dione.

Give a mechanistic explanation for the action of DPTF in such amide bond formation reactions.

1

$$Boc\text{-}Ser(Bn)\text{-}OH \xrightarrow{\text{(i)}} \underset{\text{(ii)}}{H\text{-}Val\text{-}OMe.HCl} \quad Boc\text{-}Ser(Bn)\text{-}Val\text{-}OMe$$
$$\textbf{2} \qquad\qquad\qquad\qquad \textbf{3}, 88\%$$

(i) DPTF/aqNaHCO$_3$/MeNO$_2$/RT; (ii) NEt^{i}Pr$_2$/RT

268.   A Benzothiazole from Oxidation of Mammalian Red Hair with Hydrogen Peroxide

The pheomelanins are a group of heterogeneous, reddish-brown, sulfur-containing pigments which are apparently implicated as "major determinants of the abnormal susceptibility of red-haired, fair-complexioned individuals to sunburn and skin cancer." Their structures are unknown, but 5-*S*-cysteinyldopa **1** is known to be the main biosynthetic precursor. Italian workers recently discovered that treatment of mammalian red hair with alkaline hydrogen peroxide at room temperature gave a single major product whose properties did not correspond to those of any known melanin product degradation. The same product could be obtained in up to 25% yield by degradation with 1% hydrogen peroxide in 1M NaOH of a synthetic pheomelanin sample prepared by enzymatic oxidation of **1** with tyrosinase. The structure of this new product was shown to be **2**, and it was suggested that **2** could serve as a potential marker for microanalysis of phenomelanins in pigmented tissue.

Suggest a mechanism for the formation of **2** from **1**.

$$1 \qquad\qquad 2$$

269.   Cyclopentenones from 1,3-Cyclopentanediones

It has long been known that 2,2-dialkyl-1,3-cyclopentanediones undergo smooth β-dicarbonyl cleavage to give 5-alkyl-4-oxoalkanoic acids on treatment with dilute sodium hydroxide solution. German workers have been investigating the related reactions of the triketones **1**, exploring in particular the possibilities for intramolecular aldol reactions and hence the preparation of bicyclic systems. Reaction of **1** (R = Me) with two equivalents of sodium hydroxide in water at room temperature gave an 85-95% yield of a mixture of two products, **2** (R = Me) and **3** (R = Me), in the ratio 9:1. Compound **2** (R = Me) was regarded as the "expected" product, but isolation of **3** (R = Me) was quite unexpected. This overall transformation **1** → **2** + **3** was found to be general for a variety of R = alkyl groups, but the amount of **3** increased with increasing size of R. For example, when R = n-C$_5$H$_{11}$, the yield of **3** increased to 27%.

Give a mechanistic explanation for the assumption that **2** (R = Me) was the "expected" product from reaction of **1** (R = Me) with dilute base, and suggest a mechanism for the formation of the products **3**.

| | 1 | | 2 | | 3 |

270. A New Synthesis of 1- and 2-Chloronaphthalenes by an Annulation Process

Cyclopropanes are valuable building blocks in synthesis, relief of steric strain leading to many types of ring opening reactions under thermal, reductive and oxidative conditions, and by reaction with nucleophiles and electrophiles. Japanese workers have been studying the use of *gem*-dihalocyclopropylmethanols as precursors to naphthalene derivatives suitable for elaboration into natural lignan lactones, and have discovered novel benzannulation sequences. Thus, treatment of **1** with either one equivalent of a Lewis acid (BF$_3$.OEt$_2$, SnCl$_4$, TiCl$_4$) in DCM at room temperature, or simply dissolving **1** in TFA at room temperature, led to formation of the 1-chloronaphthalenes **2** in moderate to quantitative yield. By contrast, reaction of **3** in TFA as solvent gave the 2-chloronaphthalene **4** in 78% yield.

Give mechanisms for the formation of **2** and **4** which account for the different reactivity patterns.

| | 1 | | 2 | | 3 | | 4 |

R^1 = R^2 = H, alkyl
R^3 = H, Aryl-X

271. From the Diterpene Carnosol to the Benzodiazepine Agonist Miltirone

The abundant diterpene carnosol **1** can be converted easily into the ester **2** in 90% total yield by the sequence shown in the Scheme.

Identify the intermediate **A** and suggest a mechanism for the conversion of **1** into **2**.

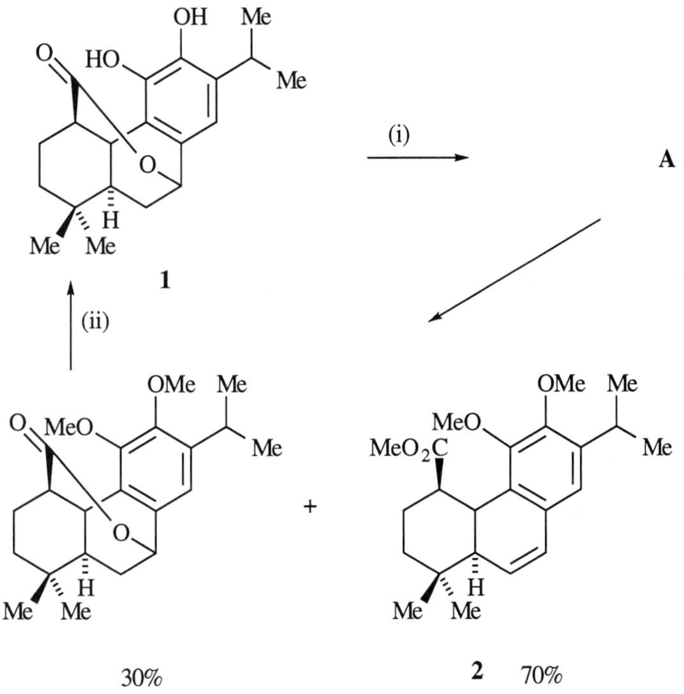

30%

2 70%

(i) MeI/K$_2$CO$_3$/Me$_2$CO/RT
(ii) BBr$_3$/CH$_2$Cl$_2$/RT

Attempts to hydrolyse the ester **2** with either KOH/H$_2$O/MeOH at 50°C or LiOH/THF/H$_2$O at reflux or AcOH in quinoline at 120°C were unsuccessful, and the starting material was recovered unchanged. Use of KO^{t}Bu/DMSO at 40-60°C for 4 hours, however, followed by aqueous acid gave a mixture of four products, the major of which (80%) was shown to be the acid **3**. Ether cleavage of **3** with BBr$_3$/CH$_2$Cl$_2$ at room temperature proceeded rapidly (5 min), and while the catechol derivative could be isolated and characterised spectroscopically, it was rapidly oxidised in air to **4**, the potent benzodiazepine agonist miltirone.

Give mechanisms for the conversion of **2** into **3** and of **3** into **4**.

3　　　　　　　　　　　　**4**

272. Synthesis of Chiral Phthalimidine Derivatives

Phthalimidines (isoindolin-1-ones) can be valuable intermediates for the synthesis of isoindoles and some natural products, and there has been recent interest in the development of simple methods for the direct conversion of *o*-phthalaldehyde into *N*-substituted phthalimidines. Condensation of *o*-phthalaldehyde with primary aliphatic amines using acetonitrile as solvent gives disappointing yields; with α-methylbenzylamine, for example, the yield of the phthalimidine **1** is only 21%. By contrast, treatment of *o*-phthalaldehyde with α-amino acids in hot acetonitrile gives generally excellent yields of the corresponding phthalimidines. With L-valine, for example, **2** is formed in 87% yield.

Suggest mechanisms for the formation of **1** and **2** which account for the higher yields observed when α-amino acids are used rather than the primary amines.

1　　　　　　　　　　　　**2**

273. Remarkable Rearrangement of a Camphor Derivative

Canadian workers have been exploring methods for the conversion of cheap and readily available camphor into enantiopure cyclopentane derivatives for use in terpene and steroid synthesis, e.g. **1** → **2**. In a recent extension of these studies, *endo*-3-bromo-4-methylcamphor **3** (0.1 mol) was treated with bromine (0.233 mol) and chlorosulfonic acid (20 ml) at 0°C for 10 minutes and then at room temperature for 18 hours. Work up with ice/saturated NaHSO$_3$ gave an 84% yield of the tribromo derivative **4**, together with minor amounts of two other products. Careful examination of this reaction by labelling studies revealed the probable pathway, for this remarkable rearrangement which, according to the researchers, "involves a combination of four Wagner-Meerwein rearrangements, two *exo*-3,2-methyl shifts, and two bromination steps."

Suggest a mechanism for the conversion of **3** into **4** which is consistent with the description given.

| 1 | 2 | 3 | 4 |

274. A Failed Attempt to Prepare Benzylidenethiophthalide by the "Obvious" Method

Belgian workers needed to prepare the benzylidenethiophthalide **1** to test its potential use as a chain-stopper for a particular polymerisation process. 2-Thiophthalide **2** was therefore condensed with benzaldehyde in THF at 0°C using potassium t-butoxide as base. Work-up consisted simply of addition of water, acidification with HCl and extraction with chloroform. None of the expected product **1** was obtained; instead, the stilbene acid **3** was formed in 52% yield.

Suggest a mechanism for the transformation of **2** into **3**.

| 1 | 2 | 3 |

275. Base-Catalysed Reactions of Highly Hindered Phenols Used as Antioxidants

During an investigation of the fate of highly hindered phenols in their performance as antioxidants, it was found that treatment of 3,5-di-t-butyl-4-hydroxybenzyl chloride with the anion of dimethyl sulfoxide gave, among other products, a colourless crystalline solid **A**, $C_{44}H_{66}O_3$. The ir spectrum (KBr) had strong bands at 3618, 1655 and 1640 cm^{-1}, while the uv spectrum (MeOH) had maxima at 231 (ε 23,000) and 272 nm (ε 5200). The mass spectrum did not reveal an M$^+$ peak, but there were intense fragments at m/z 424 and 219. The ^{1}H nmr spectrum (CDCl$_3$) consisted of six singlets at 1.10 (9H), 1.38 (18H), 2.90 (2H), 5.01 (1H, exchangeable), 6.56 (1H) and 6.81 ppm (2H) downfield from internal TMS. The proton decoupled ^{13}C nmr spectrum (CDCl$_3$) showed signals at 29.4 (6C), 30.2 (12C), 34.1 (4C), 34.6 (2C), 46.2 (1C), 47.2 (2C), 126.4 (4C), 127.7 (2C), 135.1 (4C), 145.2 (2C), 147.0 (2C), 152.5 (2C) and 186.4 (1C) ppm. Careful investigation of the reaction showed that formation of **A** was a consequence of the presence in the starting material of 2,6-di-t-butylphenol as an impurity.

Deduce the structure of **A** and give a mechanistic explanation for its formation.

276. Oxidative Rearrangement of an Aconitine Derivative

Oxonine **1**, a derivative of the diterpene alkaloid aconitine, undergoes smooth oxidation when treated with exactly one equivalent of periodic acid. The product thus obtained, which shows no aldehyde or ketone absorption in the ir spectrum, is converted into **2** when heated in dilute aqueous base with air passing through the solution.

Suggest a mechanism for the formation of **2** under these conditions.

1

2

277. Decomposition of Akuammicine

The alkaloid akuammicine **1** is known to decompose slowly when heated in methanol. In a careful study of this reaction it was found that when **1** was heated with methanol at 100°C in a sealed tube for 3 hours it had decomposed to the extent of 80%, and the betaine **2** was isolated in 70% yield. Under the same conditions, but at a temperature of 140°C, decomposition was complete within 2 hours, and the two products which were isolated were 3-ethylpyridine (48%) and 3-hydroxycarbazole **3** (60%).

Give mechanistic interpretations for these transformations.

1

2

3

278. Extending the Favorskii Reaction

Favorskii-type reactions, which are known to be sensitive to the base and solvent used and to the nature of the halogen leaving group, continue to attract attention, and in combination with other bond forming processes can often be manipulated to provide useful synthetic methods. For example, the chloroketone **1** gives pivalic acid on treatment with 40% aqueous NaOH, but only 3-hydroxy-3-methyl-2-butanone on reaction with either 20% aqueous Na_2CO_3 or 14% aqueous or ethanolic KOH. When a mixture of **1** (3.5 eq.) and benzaldehyde (1 eq.) was treated with 14% ethanolic KOH (7 eq.) at room temperature for 3 hours, however, the lactone **2** was the major product (50%), together with the unsaturated acids **3** (14%) and the alcohol **4** (5%). Use of 14% *aqueous* KOH, on the other hand, gave only the alcohol **4**, in 80% yield.

Suggest mechanisms to account for the formation of **2**, **3** and **4**.

279. Dihydroxylation/Base Treatment of the Westphalen Ketone

Dihydroxylation of the Westphalen diketone **1** with OsO_4 gives the 9,10-dihydroxy derivative, gentle treatment of which with warm methanolic KOH results in formation of **2**. Reaction of the diol with methanolic KOH under more drastic conditions (reflux, 18 hrs), however, gives **3**.

Give mechanistic interpretations for these transformations.

280. Quantitative Conversion of a Vinylogous Thioamide into a Thiophene

Addition of methyl iodide to a solution of the aminodithioacrylate **1** in acetone resulted in the rapid formation of the corresponding thionium salt in high yield. Treatment of a solution of **1** in acetone with excess of ethyl α-bromoacetate followed by excess of triethylamine, on the other hand, gave ethyl 2-methylthiothiophene-5-carboxylate in almost quantitative yield.

Suggest a mechanism for formation of the thiophene ester.

1

281. Rearrangements During Synthetic Studies on Carba-Sugars

Glycosidase inhibitors are of much current interest with respect to their potential use in chemotherapy as antidiabetic or antiviral agents. A number of carba-sugars such as valiolamine **1** are potent α-glucosidase inhibitors, and hence there has been much effort devoted to their synthesis. During a recent study by chemists in Hong Kong, which resulted in development of a facile synthesis of valiolamine and its diastereomers from (-)-quinic acid, a number of unexpected rearrangements were encountered. Thus, treatment of the alcohol **2** with triflic anhydride and pyridine in dry dichloromethane, first at 0°C then at room temperature for 3 hours, did not give any of the expected triflate. The only product isolated was the rearranged alcohol **3**, which was obtained in 76% yield. Treatment of **2** with methanesulfonyl chloride in pyridine at 0°C did give the expected mesylate, but attempted nucleophilic displacement with Bu$_4$NN$_3$ at 100°C gave only the rearranged alcohol **3**.

1   **2**   **3**

Another unexpected rearrangement was encountered in the synthesis of 1-*epi*-valiolamine, when the alkene **4** was subjected to dihydroxylation conditions. Thus, heating of a mixture of **4** with a catalytic amount of OsO_4 in a water/pyridine/t-butanol mixture containing trimethylamine *N*-oxide under reflux for 2 days gave a mixture of **5** (29%) and the α-diol **6** (14%) together with 46% of unchanged alkene **4**.

Suggest mechanisms to account for the formation of **3** and **5**.

282. A Pyrimidine to Pyrazole Transformation

A very wide range of pyrimidines is readily accessible by synthesis. By contrast, there are relatively few good general routes to imidazoles. Consequently, the pyrimidine → imidazole ring transformation can be a valuable process. Japanese workers have shown, for example, that treatment of 5-acylaminouracils of the type **1** with 5% aqueous sodium hydroxide in ethanol at reflux results in smooth conversion into imidazoles **2** in generally good yield.

Give a mechanistic interpretation of this transformation.

283. Attempted Diastereocontrol in Synthesis of a Homoallylic Alcohol

During synthetic work on the macrolide antibiotic amphotericin B, the diene **1** was required for preparation of an important building block. The key step to **1** featured diastereocontrol in the conversion of the precursor sulfone **2** into **1** by treatment first with 3 equivalents of allyllithium followed by addition of HMPA. Under these conditions **2** gave a mixture which consisted mainly of **1** (43%) and **3** (31%).

Account mechanistically for the formation of **1** and **3**.

1   **2**   **3**

284. Arsonium Ylides for Cyclopropane Synthesis

An Australian worker has described recently a novel synthesis of highly functionalised cyclopropanes which proceeds in moderate yield, and is illustrated by the following example. Addition of the 2*H*-pyran-5-carboxylate **1** to the arsonium ylide **2** in THF at 0°C gave two *trans*-diastereomers **3** in 64% yield.

Explain the formation of **3** in mechanistic terms.

1   **2**   **3**

285. Pyridoacridines from 4-Quinolones

Spanish workers have recently described a rapid method for the construction of pyridoacridines which could be of value for the synthesis of a range of cytotoxic pyridoacridine alkaloids. Thus, treatment of MEM-protected 4-quinolone **1** with 3 equivalents of LDA in THF, first at -78°C then with stirring at room temperature, followed by addition of 1 equivalent of methyl 3-formyl-2-pyridinecarboxylate gave, after quench with aqueous ammonium chloride and chromatography, a mixture of the tetracycles **2** (15%) and **3** (20%). Oxidation of either **2** or **3**, or a mixture of **2** and **3**, with aqueous CAN in acetonitrile at room temperature gave the pyridoacridine **4** in 90% yield.

Elucidate the synthetic scheme in mechanistic terms.

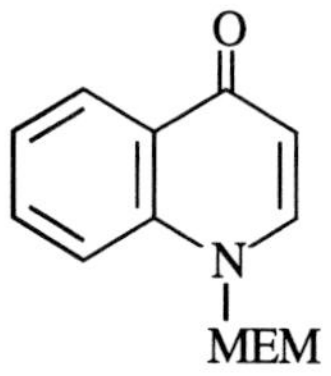

1

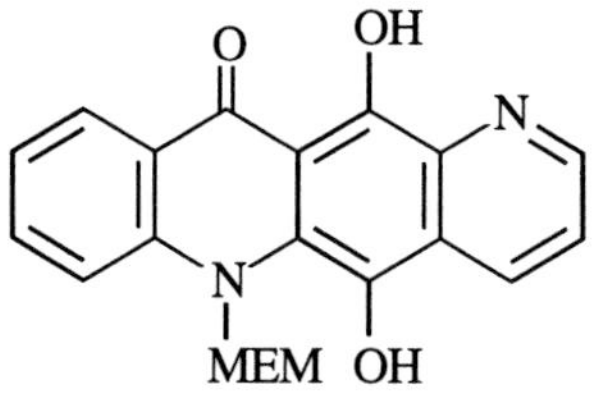

2

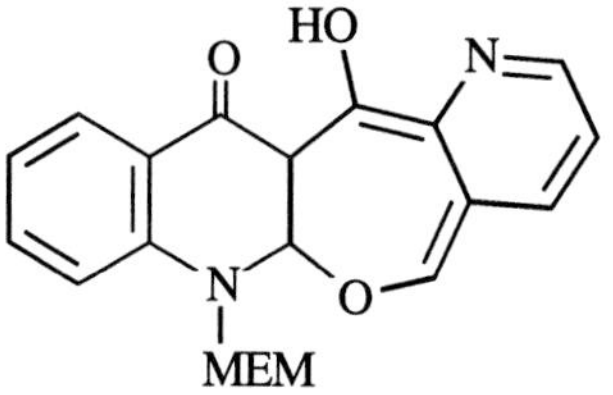

3

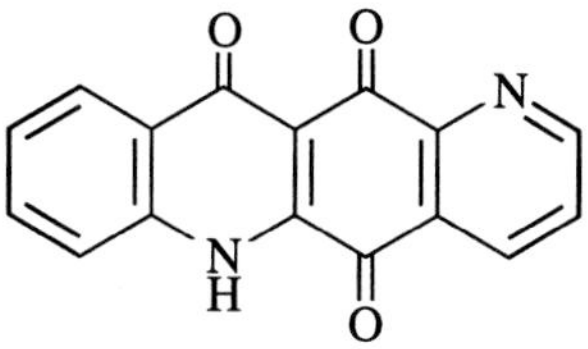

4

286. Imidoyl Radicals in a New Quinoline Synthesis

Imidoyl radicals are often prepared by the addition of carbon- or heteroatom-centred radicals to isocyanides, but can also be prepared by hydrogen atom abstraction from imines. This latter method has been adopted in a new annulation approach to quinoline synthesis. Thus, reaction of the imine **1** with phenylacetylene and di-isopropyl peroxydicarbonate in benzene at 60°C gave a mixture of the quinolines **2** and **3** in 65% yield (2:3 = 4.4).

Give mechanisms for the formation of **2** and **3**.

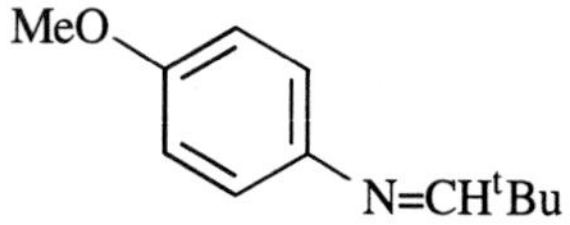

1

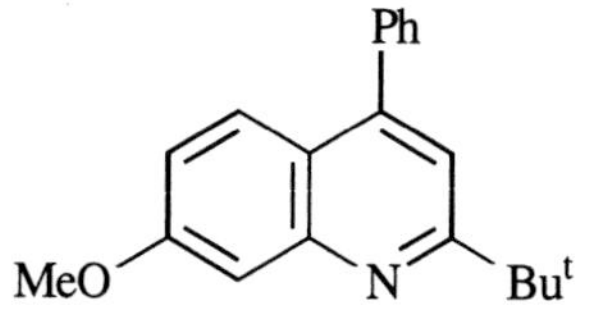

2

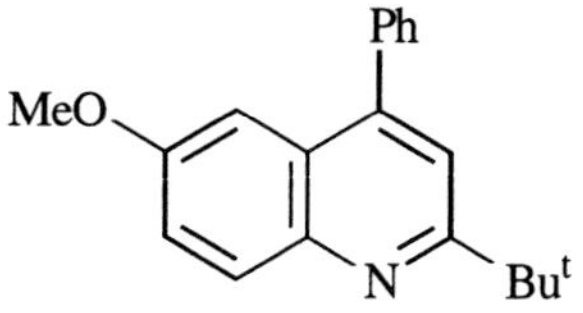

3

287. Electrolytic Fluorinative Ring Expansion Reactions

The complex $Et_3N.5HF$ is an excellent electrolyte and functions as a convenient fluorine source for the partial fluorination of aliphatic aldehydes and cycloketones. It has now been found that electrolysis of cycloalkylideneacetates **1** with $Et_3N.5HF$ at -20°C results in smooth fluorinative ring expansion to give the difluorocycloalkyl esters **2** in moderate to good yield.

Suggest a mechanism for the **1** → **2** transformation.

1, R = H or CO_2Et **2**, R = H or CO_2Et
    n = 0, 1 or 2

288. An Isoxazole to 1,3-Oxazine Transformation

It was reported in 1977 that reaction of 3-phenyl-5-isoxazolone with excess of $POCl_3$ in DMF under Vilsmeier conditions gave the isoxazoline **1**. Treatment of **1** with MeOH/NaOMe and with 10% NaOH solution was claimed to give the aldehyde **2** and the azirine **3** respectively.

         1   **2**   **3**

Subsequent attempts to repeat this work failed. Instead, reaction of 3-phenyl-5-isoxazolone with $POCl_3$/DMF, first at room temperature then for 30 minutes at 60°C, gave the isoxazolone **4** in 35% yield after standard aqueous quench and neutralisation with 5% $NaHCO_3$ solution. When longer reaction times were employed, or when the reaction temperature was raised to 80°C, the product, isolated in 71% yield, was the 1,3-oxazinone **5**, and **4** was shown to be the precursor to **5**.

Suggest a mechanism to account for the formation of **5**.

4   5

289. Synthesis of Folate Antimetabolites : A Furan to Pyrrole Transformation

Reaction of *o*-aminonitriles with guanidine is a widely used and efficient method for the preparation of condensed 2,4-diaminopyrimidines, **1** → **2**. This procedure was adopted with a variety of 2-amino-3-cyanofurans, in the expectation that the products would be furo[2,3-*d*]-

Ar or HetAr
1   2

pyrimidines, which were required for biological evaluation as analogues of the corresponding pyrrolo[2,3-*d*]pyrimidines. There is intense current interest in these latter compounds as folate antimetabolites.

The reactions with the aminocyanofurans produced an unexpected result, however, as illustrated by the following example. 2-Amino-3-cyano-4-methylfuran was added to a solution of guanidine free base in ethanol, and the mixture was heated under reflux for 24 hours. Standard work-up gave the pyrrolo[2,3-*d*]pyrimidine **3** in 67% yield, and *not* the expected furo[2,3-*d*]pyrimidine.

Give a mechanistic explanation for the formation of **3**.

3

290. Rearrangement of a Silylacetylenic Ketone

An unexpected rearrangement was observed during studies directed towards the synthesis of bicyclic keto silanes by the thermal rearrangement of cycloalkanones bearing an ω-silylacetylenic chain β to the carbonyl group. Thus, the bicyclic enol silyl ether **2** was formed in 60-65% yield when the ketone **1** was heated neat at 300°C for 2 hours.

Give a mechanism for this transformation.

291. And More Silyl Rearrangements : A Brook-Retro-Brook Sequence

Stereopure epoxide **1** was prepared and treated with 3.6 equivalents of t-butyllithium in THF/HMPA at -78°C. The intention was that formation of the anion at the benzylic carbon would lead to a 4-*exo*-epoxide ring opening reaction; a subsequent [1,2]-silyl shift (Brook rearrangement) would generate the oxetane **2** with stereocontrol at all three stereocentres. Anion formation proceeded smoothly at -78°C, then 1 ml of 1 M hydrochloric acid was added and the product isolated. Obtained pure in 40% yield, this was shown to be the aldehyde **3**. No oxetane **2** was obtained.

Suggest a mechanism to account for the conversion of **1** into **3**.

292. Two Syntheses of Dehydrorotenone

Dehydrorotenone **1** can be readily transformed into rotenone **2** by reduction of the chromone to the chromanol followed by Oppenauer oxidation. Two elegant syntheses of dehydrorotenone **1** used DCC as a key reagent as follows : (i) treatment of derrisic acid **3** with DCC/Et$_3$N followed by reaction of the intermediate thus obtained with sodium propanoate in ethanol gave **1**. (ii) Condensation of tubaic acid **4** with the pyrrolidine enamine derived from **5** in the presence of DCC also gave **1**.

Elucidate these syntheses of **1** and explain in mechanistic terms the role of the DCC.

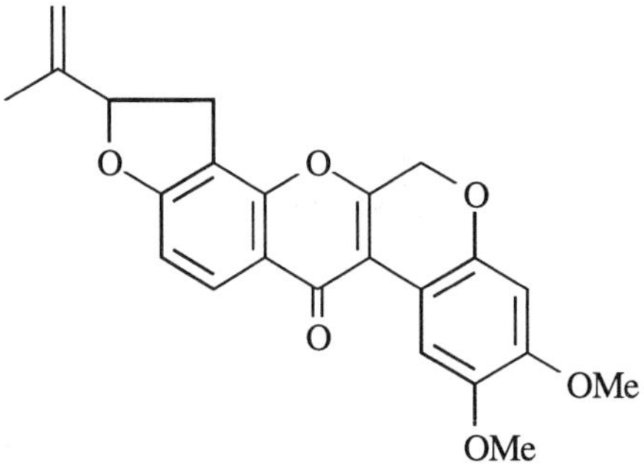

1

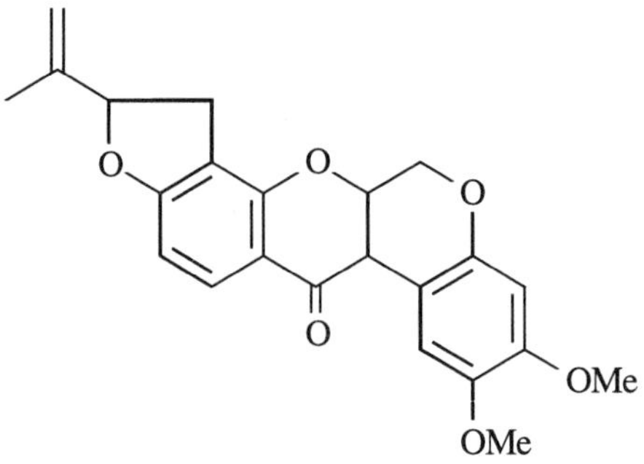

2

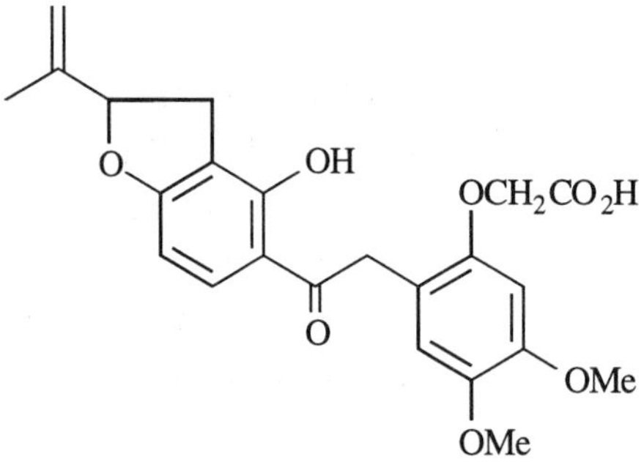

3

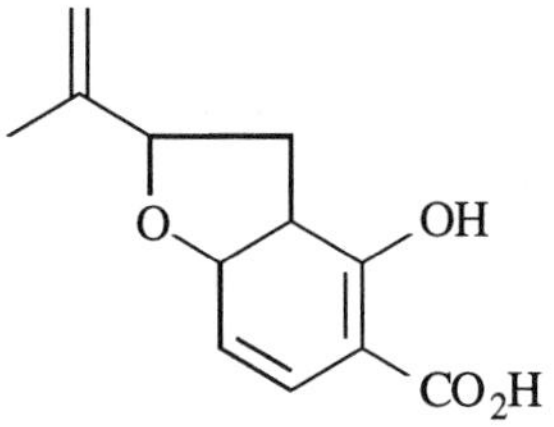

4

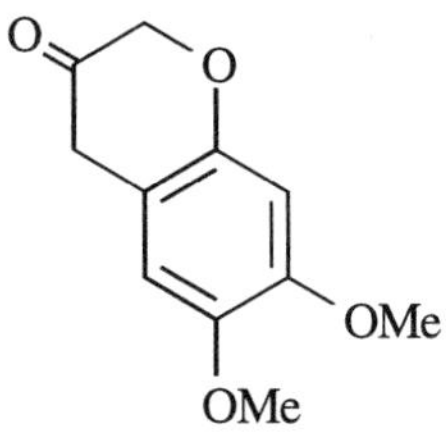

5

293. Radical-Induced Decarboxylation of a Lactone

Assessment of drug stability towards hydrolysis and radical-induced decomposition is an important preformulation exercise. Given the relative complexity of many modern pharmaceuticals, unexpected results from such studies are not uncommon, as illustrated by the following example. The antibiotic frenolicin-B, **1**, was heated with AIBN in methanol in the presence of air for two days. This gave a number of "uncharacterizable materials" together with one major product which was shown to be the racemic pyranonaphthoquinone **2**.

Suggest a mechanism, or mechanisms, for the radical-induced degradation of **1** into **2**.

294. Aryl Azide Thermolysis : "A Series of Rather Involved Rearrangement Reactions"

In 1996, Indian workers reported the formation of a most unusual product from the thermolysis of a highly substituted aryl azide (S.V. Eswaran, H.Y. Neela, S. Ramakumar and M.A. Viswamitra, *J. Heterocyclic Chem.*, **33**, 1333-1337 (1996)). Thus, heating of the azide **1** in chlorobenzene for 4 hours at 130°C followed by removal of the solvent by distillation under reduced pressure and chromatography of the residue on silica gel gave the product **2** in 19% yield. The structure of **2** was established by X-ray analysis, but no mechanism was suggested for its formation.

Devise a mechanism to explain the formation of **2** from **1**.

295.   An Efficient Route to Hexakis(trifluoromethyl)cyclopentadiene

An efficient synthesis of hexakis(trifluoromethyl)cyclopentadiene **1** has been described very recently. Thus, a mixture of 1,1,3,3,3-pentafluoropropene (2 eq.), perfluoro-3,4-dimethylhexa-2,4-diene (1 eq.) and dried cesium fluoride (4 eq.) in anhydrous acetonitrile was sealed in a Carius tube which was rotated, using a mechanical arm, at room temperature for 48 hours. This gave **1** in 74% yield.

Suggest a mechanism for the formation of **1**.

1

296.   Exploitation of the Boulton-Katritzky Rearrangement: Synthesis of 4,4'-Diamino-3,3'-bifurazan

Aminofurazans are attracting increasing attention because of their range of biological activities and their potential uses as "energetic materials". Russian workers have described recently a simple, one-pot method for the preparation of 4,4'-diamino-3,3'-bifurazan **1** which consists of slow addition of an aqueous solution of potassium hydroxide to a suspension of 3,4-di(hydroximinomethyl)furoxan **2** and a salt of hydroxylamine in aqueous DMSO with stirring and heating. This gives **1** in 18% yield, which is much better than the yield obtained by a previous, multistep synthesis (3% overall). Moreover, the furoxan **2** is readily available from nitromethane in two steps in 78% yield.

Give a mechanism for the formation of **1** from **2**.

1   **2**

297. Rearrangement During Hydrolysis of a Cyclohexadienone

The diterpenoid **1** from the heartwood of *Callitris macleayana* is the Diels-Alder dimer of the dienone **2a**. The acetate **2b** of the alcohol **2a** is readily available by oxidation of 5-isopropyl-2-methylphenol with lead tetraacetate, but all attempts to hydrolyse **2b** to **2a** failed. "Dimeric indans" were obtained under acidic conditions, while use of potassium hydroxide in methanol at room temperature for 15 minutes followed by acidification with 1M hydrochloric acid, extraction and repeated chromatography over silica gave the three products **3**, **4** and **5** in the relative distribution 11.7, 63.7 and 24.7%. The same products were formed in similar proportions when the reaction temperature was varied from 0°C to 64°C; treatment of any of the products **3**, **4** or **5** with potassium hydroxide in methanol also gave a mixture of **3**, **4** and **5**.

Suggest mechanisms to account for the formation of **3**, **4** and **5**.

1

2a, R = H
b, R = Ac

3

4, R^1 = Me, R^2 = OH
5, R^1 = OH, R^2 = Me

298. Less Common Sigmatropic Rearrangements : [3,4], [3,5] or even [2,3,2]?

A small number of natural products are known based on the 1,2-dithiin system, e.g. thiarubrine A, **1**. They occur in plants used in traditional medicine to treat skin infections, intestinal parasites and fevers. The 1,2-dithiins are rather unstable, 8π-electron antiaromatic systems, and any synthetic route almost inevitably requires formation of the heterocyclic ring as the final step. American scientists studied the thionation of constrained 1,4-diketones of the type **2** in the hope that low temperature oxidation of the dienedithiol tautomers of the thioketones thus produced would lead to the fused 1,2-dithiins. Thionation apparently proceeded as expected, but the products which were isolated were unexpected, and were formed by rearrangement. Thus, treatment of the *endo,exo*-diketone **2** (Ar = Ph) with bis(tricyclohexyltin) sulfide, [(C₆H₁₁)₃Sn]₂S, gave a mixture of **3** (45%) and **4** (20%), while the *endo,endo*-diketone **2** (Ar = Ph) gave only **4** in 75% yield (no experimental conditions were given for the reactions).

Suggest mechanisms to account for the formation of **3** and **4**.

1 Thiarubrine A   2

3   4

299. Failure to Construct an Oxetane by S$_N'$ Rearrangement

A pentacyclic diterpene **1** called dictyoxetane contains a most unusual subunit, a 2,7-dioxatricyclo[4.2.1.0³,⁸]nonane. During model studies designed to provide access to this key subunit the bicyclic ether **2** was synthesised in the hope that S$_N'$ displacement would generate the unsaturated tricyclic oxetane. There was no reaction when **2** was treated with base. Reaction with a catalytic amount of *p*-toluenesulfonic acid in DMF at 75°C for 24 hours resulted only in formation of 4-methylacetophenone. The hydroxy mesylate **2** is also reported to decompose to 4-methylacetophenone on storage.

Suggest a mechanism to account for the formation of 4-methylacetophenone from **2**.

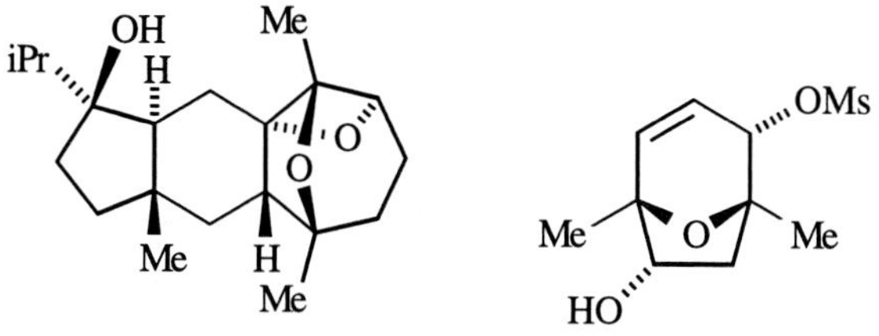

1

2

300. Acid-Catalysed Isomerisation of a Tetraspiroketone

German chemists synthesised the tetraspiroketone **1** and carried out theoretical calculations in an attempt to decide whether acid-catalysed isomerisation would lead preferentially or exclusively to the bispropellanone **2** or to the pentacyclic ketone **3**. Force field calculations failed to reveal any significant preference for rearrangement, and hence experimental clarification was sought. The ketone **1** was unstable to acid, and was quantitatively isomerised to **3** within 30 minutes at 80°C when treated with one equivalent of a 0.25 M solution of anhydrous *p*-toluenesulfonic acid in benzene.

Outline mechanisms by which **1** might be expected to undergo acid-catalysed isomerisation to **2** and/or **3**.

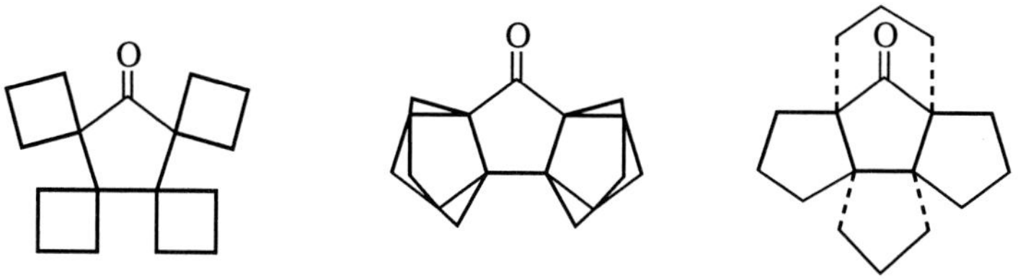

1   **2**   **3**